D0984533

On Faith and Science

On Faith and Science

EDWARD J. LARSON
AND
MICHAEL RUSE

Yale
UNIVERSITY PRESS
New Haven and London

Published with assistance from the foundation established in memory of Calvin Chapin of the Class of 1788, Yale College.

Yale University Press books may be purchased in quantity for educational, business, or promotional use. For information, please e-mail sales.press@yale.edu (U.S. office) or sales@yaleup.co.uk (U.K. office).

Set in Janson Roman type by Tseng Information Systems, Inc., Durham, North Carolina.
Printed in the United States of America.

ISBN 978-0-300-21617-2 (hardcover : alk. paper)
Library of Congress Control Number: 2017932272
A catalogue record for this book is available from the British Library.

This paper meets the requirements of ANSI/NISO Z39.48-1992 (Permanence of Paper).

10 9 8 7 6 5 4 3 2 1

EJL:

To my mentor in science and religion,

Ron Numbers

MR:

To my colleague in science and religion,

Bob Richards

Contents

On Faith and Science

Introduction: What's the Fuss?

In his short story "The Duniazát" from 2015, Salman Rushdie, born into a Muslim family in British India but targeted for death for his writings by such prominent Islamic clerics as Iran's Supreme Leader Ayatollah Khomeini, relates the story of the twelfth-century Muslim philosopher, physician, and jurist Ibn Rushd. Through the influence of his commentaries, Ibn Rushd (known in the West as Averroës) helped to introduce Aristotelian philosophy to medieval Europe and thus set the path of Western thought on a course that, through many contingent twists and turns, ultimately generated modern science. In Rushdie's account, Ibn Rushd never intended this result but facilitated it nevertheless, and it made him a heroic figure. As "The Duniazát" makes clear, Ibn Rushd suffered for his "science" at the hands of Islamic clerics some eight hundred years ago, much as Rushdie currently suffers at those hands for his writings, and in this sense the story introduces both the historical significance and ongoing relevance of our topic.

"In the year 1195," Rushdie's account begins, "the great philosopher Ibn Rushd, once the *qadi*, or judge, of Seville and most recently the personal physician to the caliph Abu Yusuf Yaqub in his home town of Córdoba, was formally discredited and disgraced on account of his liberal ideas, which were unacceptable to the increasingly powerful Berber fanatics who were spreading like pestilence across Arab Spain." By this point in his life, Ibn Rushd had written works on philosophy, including various branches of "natural philosophy" (as the sciences were then known), and commentaries on many of the known works of Aristotle. Perhaps his most important work, "The Incoherence of Incoherence," challenged the then century-old claims of Muslim theologian Ghazali of Tus, who in the work "The Incoherence of Philosophy" argued against Aristotelian notions of natural law. "Philosophy believed in the inevitability of cause and effect," is how Rushdie summaries Ghazali's position, "which was an insult to the power of God, who could easily intervene to make causes ineffectual and alter effects if He so chose." Ibn Rushd's defense of divinely created natural law led to his banishment in an allegedly classic case of the interference of religion with science.

Rushdie, whose own family name was adopted by his father in tribute to Ibn Rushd, repeats the famous example of the combustibility of cotton that appears in the battling treatises of the two medieval Islamic scholars. Cotton catches fire when licked by a flame, Rushdie writes in explaining Ibn

Rushd's Aristotelian position: "It's how things are. The law of nature. . . . Causes have their effects." Not so, he counters in relating Ghazali's religious view: "The cotton caught fire because God made it do so, because in God's universe the only law is what God wills." For challenging the sovereignty of God, Ibn Rushd is banished from the court when his patron, the caliph Yaqub, becomes vulnerable to religious extremists within his realm. "He used words that many of his contemporaries found shocking, including 'reason,' 'logic,' and 'science,' which were the three pillars of this thought, the ideas that had led to his books' being burned," Rushdie explains in a clear allusion to the ongoing state of science in the modern Islamic world. Although Rushdie might not get every element of this historic encounter exactly right, he captures the basic narrative as it has come down to us today and still lives in the Islamic world.

In Rushdie's fictional version of what follows, Ibn Rushd's lover probes the consequences of both lines of reasoning. "So anything can happen if God decides it's O.K. . . . A man's feet might no longer touch the ground, for example. He could start walking on air." Ibn Rushd concedes this would be a miracle of the type people should expect in a world governed by Ghazali's theology: "God changing the rules by which He chooses to play." But then she counters in a logical extension of Aristotelian reasoning, "Suppose I suppose . . . that God does not exist. Suppose you make me suppose that 'reason,' 'logic,' and 'science' possess a magic that makes God unnecessary." This was Ghazali's concern, but Ibn Rushd in both his writings and Rushdie's account dismisses it. "That really would be a stupid supposition," Rushdie has Ibn Rushd declare. Quite to the contrary, Ibn Rushd argues that the certainty of knowable cause-and-effect relationships in nature proves that a rational Creator *does* exist and that the benevolent results of those natural laws testify to God's goodness. Ibn Rushd maintains that a religious worldview enables science and that the findings of science support religion. The two ways of knowing complement each other.

Although not mentioned in Rushdie's short story, which is directed against Islamic restrictions on reason, concerns like those expressed by Ghazali also arose in Christian Europe when Ibn Rushd's Aristotelian commentaries arrived there a century later. In 1277, for example, the Catholic bishop of Paris condemned various Aristotelian teachings rooted in natural law. In Western Christendom, church-imposed limits on certitude over cause and effect stimulated speculative thought and spurred the empiricism that some historians see as ultimately advancing science. Nonetheless, those underlying concerns remain, resurfacing even today in some religious objections to the theory of evolution. If God is seen as specially creating every kind of living thing, then people can scarcely question God's existence or character. In contrast, if the vast array of living things, including humans, evolved in series of natural cause-and-effect steps all the way back to space dust, then it is easy to dismiss God's relevance.

After his patron wins a great victory over the Christian king of Castile in 1197, Ibn Rushd is rehabilitated. The victory enables the caliph to overcome religious extremists within his realm, but in the long run, Rushdie's story reminds us, the zealots gain sway in many parts of the Islamic world to the detriment of science. Ibn Rushd's Aristotelian commentaries instead take root in the Christian West, leading to both the type of science fostered by belief in a rational Creator knowable through creation and the type of science that pushes God so far back in the cause and effect of nature as to make a Creator all but irrelevant. "Perhaps, as a godly man, Ibn Rushd would not have been delighted by the place history gave him," Rushdie concludes, "for it is a strange fate for a believer to become the inspiration of ideas that have no need of belief, and a stranger fate still for a man's philosophy to be victorious beyond the frontiers of his own world but vanquished within those borders, because in the world he knew it was the children of his dead adversary, Ghazali, who multiplied and inherited the kingdom."

FROM IBN RUSHD TO GALILEO

The story of Ibn Rushd's banishment offers an Islamic version of the West's well-known Galileo affair, which has served over the past 350 years as "Exhibit A" for the seductive and tenacious belief that the past and present relationship between science and religion is best characterized by conflict. For holding on scientific grounds that Earth orbits the sun, in contradiction to a supposed biblical view of a geocentric universe, which seventeenth-century accounts maintained, the Italian scholar Galileo Galilei was put under house arrest by the Catholic Church, forced to recant, and barred from publishing his astronomical findings. While recent historical research into the affair has shown it not to have been a simple case of religion versus science, the so-called conflict thesis that it supports remains alive and well in many minds even though, more than three centuries after it happened, the Catholic Church apologized for its role in the matter. In this book, we argue that the relationship between science and religion is more complex than any simple notion of either conflict or complementarity allows. The persistence of more simplistic views, however, is the reason why we think this book is needed.

In speaking of the conflict thesis, we're not talking about an actual war between science and religion, as if mental constructs can fight like countries do, marshaling armies and taking territory. We're talking about how people perceive the relationship between science and religion or, perhaps, what goes on in an individual's mind when confronted with the claims of science and religion. Conflict is a metaphor, of course, but the recognition that its application to the relationship between science and religion is both seductive and tenacious suggests that many people respond to it and see it as valid.

Metaphors are thrown around all the time, but they only survive if they resonate with people. For example, an American sports promoter once referred to the annual college basketball championship tournament as "March Madness," and the metaphor stuck. The tournament does not actually drive basketball fans "mad" in the clinical sense of that term, but the metaphor seemed apt to those who follow the sport. Once the phrase took hold, it helped to feed the "madness." Similarly, some nineteenth-century partisans spoke of "conflict" or "warfare" between science and religion, and that characterization seemed apt to enough people that it stuck, and thereafter probably reinforced the sense of discord.

The historiography of the Galileo affair illustrates this point. In the seventeenth century, Protestants used the episode to club Catholicism. Enlightenment secularists turned it on all Christianity during the eighteenth century. And in the nineteenth century, it applied against religion generally. This can be seen in two widely popular and highly influential American histories of the "conflict" written during the nineteenth century—one by John William Draper and one by Andrew Dickson White. Draper inexplicably exonerated the Orthodox Christian Church from criticism on this score but, as leading historians from that faith acknowledge, any lack of conflict between Galilean science and the Eastern church likely came from the failure of that science to reach Orthodox realms. Unlike Western Christianity, which sees God's hand in creation and stresses the religious word becoming physical flesh, Eastern Christianity with its unworldly spirituality did not provide fertile ground to raise issues of science versus religion because science never found footing there. Rushdie's story about Ibn Rushd's banishment fits the conflict model and thereby resonates with widely held views about science in some Islamic countries, where teaching about human evolution and even Copernican astronomy are restricted.

In the nineteenth century, however, not every American or European historian followed the lines taken by Draper and White. Other nineteenth-century histories presented religion as fostering science—with some Protestant historians claiming that the Reformation jump-started science and some Catholic historians praising Jesuit support for science. Thus, as history, the works of Draper and White were not fully representative even in their day, much less today, since the previously dominant view of religion complementing science still has its vocal proponents, perhaps most notably the Australian historian of science Peter Harrison. Harrison presents a nuanced view of how seventeenth-century radical Protestant theology drove the rise of empiricism in British science and made it popular. In short, he and others stress that science was a distinctly Christian and particularly Protestant development, though one that is now largely unmoored from its religious anchor. And the various nineteenth-century interpretations of the relationship between science and religion—from conflict to collaboration—did not solely appear in works of history. Scientists expressed these views, too.

The conflict thesis appeared in the writings of such prominent nineteenth-century scientists as Charles Lyell, Charles Darwin, T. H. Huxley, and Francis Galton, who are featured in chapters 3, 4, 5, and 6, respectively, of this book. These British scholars typically focused their fury on Catholicism (often invoking the Galileo affair), but other religions suffered their assaults as well. Huxley made this explicit in his published essays *Science and Christian Tradition*, where Naturalism can be equated with empirical science and Supernaturalism with revealed religion. "From the earliest times of which we have any knowledge, Naturalism and Supernaturalism have consciously, or unconsciously, competed and struggled with one another," he wrote. Protestants joined with secularists in using science to debunk Catholicism, Huxley noted, but "their alliance was bound to be of short duration, and, sooner or later, to be replaced by internecine warfare." Here is the conflict thesis expressed by a combatant in the fray—and so it should carry considerable evidentiary weight. Huxley, Darwin, Galton, and Lyell were all partisans in the public debate over the scientific theory of evolution, however, which excited more religious opposition than other nineteenth-century scientific ideas.

At the same time, other leading British scientists were expressing very different views of the relationship between science and religion. Think of Lord Kelvin, James Clark Maxwell, or Michael Faraday, for instance, all of whom are discussed in chapter 2. These physicists held at least as much stature as Lyell, Darwin, and Huxley in the world of nineteenth-century science, yet none of them declared war on religion. Quite to the contrary, they saw their science and religion as complementing each other. In short, although the conflict thesis was alive and well in nineteenth-century European thought, so was the opposite view.

COMPLEXITY AND BEYOND

A similar diversity of opinion on the relationship of science and religion existed during the twentieth century and continues into the twenty-first century. At least in the United States, the period has witnessed a continuing divide between science and religion, with both flourishing in their separate spheres. Housed in ever expanding research universities and fueled by unprecedented amounts of public funding, American science has assumed global leadership in terms of the sheer number of peer-reviewed articles published and Nobel Prizes received in virtually every scientific discipline. The technological payoff has transformed American industry, agriculture, and warfare. The world (or at least much of it) has taken note. Not only in the United States, but in China, India, and Europe, the study of science, technology, engineering, and mathematics (known by the acronym STEM) has exploded among top university students, with a commensurate retreat of the humanities. At the same time, surveys by Gallop, Pew, and other polling organizations have found that a greater percentage of Americans regularly attend religious services and profess belief in God than the people of any other scientifically advanced nation. Yet surveys by James Leuba, Elaine Ecklund, and the historian co-author of this book also suggest that these percentages drop off precipitously for American scientists—particularly at the higher echelons of the profession—and Americans in their twenties, especially STEM students and young professionals. This provides a context in which some conservative Christians can denounce objectionable scientific theories as the work of "atheistic scientists." And our informal survey of STEM students at UCLA—a world-class university in that field—found virtually all of them believing that there was an inevitable conflict between science and religion. "There is some interaction," one student added to what was designed as a yes-no question, "and the interaction is purely divisive." Another tried to explain: "There is always tension betwixt science and preexisting structures that are resistant to change." Every student who answered no also qualified the answer by noting that the conflict, although common, was not strictly inevitable.

Perhaps the most significant development in the relationship of science and American religion over the past century has been the disengagement of mainline Protestantism from the science-religion dialogue. In the wake of William Paley's popular works of natural theology, mainline Anglo-American Protestants regularly invoked science in support of their religious beliefs during the nineteenth century and sought to reconcile science with religion. In marked contrast, such preeminent twentieth-century Protestant theologians as Karl Barth, Paul Tillich, and Reinhold Niebuhr virtually ignored science in their theological writings. Mainline Protestants joined most Catholics in largely reserving their comments about science to ethical issues raised by technological applications of science (such as eugenics or nuclear weapons) and to making the general observation that modern scientific theories (such as the Big Bang and quantum indeterminacy) still leave room for God.

During the twentieth century, however, evangelical, fundamentalist, and Pentecostal or charismatic churches displaced mainline ones as the center of gravity within American Protestantism. Their influence has also grown in much of the developing world, particularly in sub-Saharan Africa, South America, the Pacific Islands, and parts of Asia. Many in these churches feel that their beliefs are under siege from scientists—particularly from Darwinists but also from some physicists, neurobiologists, and psychologists—and some church leaders have lashed out against these threatening ideas. In this context, American Christians periodically have stirred mass movements against the theory of organic evolution over the past century. Presbyterian politician William Jennings Bryan did so in the 1920s, resulting in legal limits on the teaching of evolution in some public schools, leading to the 1925 trial of high school teacher John Scopes for violating one such law in Tennessee. During the second half of the twentieth century, Baptist engineering professor Henry Morris and Australian-born evangelical apologist Ken Ham helped to revive a literal reading of the Genesis account of creation among conservative Protestants. These readings included the idea that the universe, the earth, and each kind of living thing was specially created by God within the past ten thousand years, and this literalism has prompted widespread demands for teaching this so-called creation science alongside evolution in American public schools. In attacking mainstream modern scientific theories in biology, astronomy, physics, geology, paleontology, and psychology, however, these religious critics typically claim they are defending a pure form of science that looks only to facts in nature and admits the possibility of supernatural causation. So for them, it is a war against modern methods of doing science rather than science itself. Still, scientists take the brunt of it.

Each of these episodes has breathed new life into the conflict thesis and evoked comparisons to Galileo's persecution. Scopes's defenders frequently compared the young Tennessee schoolteacher to Galileo, for example, with his defense counsel quoted in the widely reprinted trial transcript as declaring in court, "Every scientific discovery or new invention has been met by the opposition of people like those behind this prosecution who have pretended that man's inventive genius was contrary to Christianity." The lead prosecutor responded, "They say it is a battle between religion and science, and in the name of God, I stand with religion." Similarly, Morris and Ham have presented Darwinian science as at war with Christianity, and some scientists have responded in kind. In *The Blind Watchmaker*, for example, British biologist Richard Dawkins takes aim at "redneck" creationists and "their disturbingly successful fight to subvert American education and textbook publishing." His books, particularly the 2006 screed *The God Delusion*, which has sold over three million copies worldwide, clearly touched a nerve that has also been tapped during the early twenty-first century by philosopher of science Daniel Dennett, neuroscientist Sam Harris, and journalist Christopher Hitchens, who are sometimes collectively known as the "New Atheists." They view religion as the root of all evil and, together with such allies as Nobel Prize–winning physicist Steven Weinberg, biologist Jerry Coyne, and linguist Steven Pinker, use science to debunk it.

Further, as in Galileo's day, the weapons of this warfare are not limited to words. In 2015, the Bangladeshi-American software architect and science blogger Avijit Roy, who like Dawkins compared religion to a virus and blogged that "the vaccine against religion is to build up the scientific approach," was hacked to death by a mob in his hometown of Dhaka, Bangladesh, while police stood nearby. Two years earlier, Roy's name had appeared on a list of eighty-four atheist bloggers sent to local newspapers, and after his murder an Islamic militant group claimed credit for the attack.

The conflict model still survives among historians and philosophers of science, much as it does in popular culture. However, some scholars, encouraged by the likes of historians Ronald Numbers and John Brooke, have begun reassessing the historical relations between science and religion, shifting to a middle-ground "complexity thesis." The authors of this book (and by "we" or "us" in this book, we mean the authors), as representatives of both disciplines—Michael Ruse was trained as a philosopher of science and Edward Larson as a historian of science—generally agree with this shift but fear that it too has been pigeonholed. The overall relationship between science and religion is simply *too* complex to categorize as always conflicting or always complementary. It can be both, either, or neither depending on the situation and context. But since it can be either as well as neither, it is not inevitably complex. It is often seen by some engaged with it, such as Avijit Roy and his attackers, as straightforward warfare. As emblematic of the complexity-thesis approach, some proponents point to their reinterpretation of the Galileo affair as more of an unfortunate battle of wills between an insecure new pope and a stubborn old scientist, leading to a benign house arrest, than as a clash of medieval Catholicism and modern science resulting in an unconscionably harsh and humiliating sentence. In contrast, we see conflicting worldviews setting up the episode and giving it lasting meaning. And frankly we see no excuse for how the church acted. Thus we favor what might be called a "coexistence" approach, which views religion and science as two big messy and sometimes internally inconsistent categories of human perception and understanding that coexist in the same place and time, sometimes in a complementary or conflicting relationship but most often in a complex one, with both categories currently growing in influence and authority in many regions.

In short, given the shrill tone of the ongoing controversy over creation and evolution in the United States, Africa, the

Pacific Islands, and elsewhere, coupled with the ongoing global influence of both secular scientism and sectarian religion—as reflected in the attack on Roy and the popularity of Dawkins's books—we cannot wholly dismiss the conflict metaphor as an antiquated relic. To some partisans on both sides of these controversies, and even among a cross section of beginning STEM students at UCLA, conflict remains how they see the relationship between science and religion. Yet we can readily cite devoutly religious evolutionary biologists, such as geneticist Francis Collins, the discoverer of the cystic fibrosis gene and head of the U.S. National Institutes of Health, who characterizes himself as an evangelical Protestant and theistic evolutionist. Some recent cases suggest even closer complementarity between science and religion than shown by Collins, such as the example of noted American physicist Charles Townes, who received the Nobel Prize for his role in inventing the laser. "Religion is aimed at understanding the purpose and meaning of our universe, including our own lives," Townes said in receiving the Templeton Prize in 2005. "If the universe has a purpose or meaning, this must be reflected in its structure and functioning, and hence in science." Given the diversity of views on the topic, much remains to explore in the relationships between science and religion. Taking a historical perspective informed by the philosophy of science, we move field by field in the book alternating chapters in which our philosopher and our historian each took the lead in drafting but for which both fully participated. In doing so, we hope to raise the questions that continue to occupy scientists, theologians, philosophers, historians, and the general public. To us, it is the story of science *and* religion, not science *or* religion. With our historian having drafted this introduction and our philosopher up next, we invite readers to join us in our dialogue and, by doing so, make it into a conversation.

CHAPTER ONE

Looking Up to God or the Cosmos

IMAGINE people living in the Mediterranean region or the Middle East about three millennia ago. They might be peasants or farmers or those who look after flocks, or they could be engaged in a trade, making clothes or tools or houses. If they were Greek or Phoenician, they might be sailors. What are the things that govern their lives—the natural things, that is, not the human powers that rule them? Most obviously, the changing patterns of the day—sun up follows sun down follows sun up. Of course, this is hardly something people today are unaware of, but think how much more dramatic and prominent it would have been back then. When it starts to get dark, most people today turn on the lights and keep going about their business. This isn't to say that people are no longer aware of the stars and the moon and so forth, but for most of those living in modern times—especially in towns and cities—these are no longer dominant aspects of their lives.

Back then, three thousand or more years ago, things would have been very different. Night falls, the stars appear, and the moon shines or doesn't, and that shapes people's world until daybreak. No streetlights, no neon signs, no automobiles rushing along the roads. Human activities slow down or stop. Everything becomes very dark except for what people see up above them in the heavens. And what they see is going to be a familiar world, and they make out certain patterns in the night sky. The stars stay in the same relation to one another night after night, year after year—so much that people see recognizable shapes in their positions, give names to certain clusters or constellations, and might speculate that these have the power to influence earthly events. These constellations move across the sky in a dependable, regular arc during the night, rising in the east and setting in the west, with those passing directly overhead making up the zodiac. But not everything in the night sky is so dependable.

The moon, for instance, is by far the brightest light—so bright that people can travel and farm by it when it is full. Although it too goes around the sky each night, it does not stay in the same position relative to the stars from one night to the next but drifts steadily backward among the constellations of the zodiac, moving around to its own rhythm or pattern, and sometimes shining out fully and sometimes vanishing into a fine crescent or less.

Then there are the so-called wandering stars, which many early peoples named after gods, or thought were gods—Mercury, Venus, Mars, Jupiter, and Saturn, for example. Like the fixed stars, these wandering stars also circle across the sky from the eastern horizon to the western one every night, but from night to night they change their position among the fixed stars in a wild pattern that defies easy explanation. While remaining in the zodiac, when plotted over time against the background stars, these wanderers appear to loop the loop, stopping in their paths, moving backward, and then stopping again and once more regaining their forward-moving trajectory. All pretty complex, but not so complex that people could make no sense of the patterns, and in some societies, the ancient Babylonians particularly, there was accurate mapping and sophisticated calculating of the motions and predictions of what might be expected to happen and when. The Babylonians were especially careful about this because they viewed the wanderings stars as gods, or at least the paths of gods on the vault of heaven. By studying their movements among the constellations, they believed that one could better understand how those gods impact us. This study became known as astrology, and for centuries so many people placed such great faith in it that the casting of astrological horoscopes financed most work in astronomy. Well into the 1600s, such heroes of modern science as Kepler and Galileo pursued this practice—an example of religious beliefs fostering scientific research.

THE TWO-SPHERE MODEL

It was the ancient Greeks, beginning around 500 BCE, who started to put it all together and to make rational sense of things. With them, science and philosophy began to emerge. They realized that rather than stop at just description, people could move on to make rational sense of things—and that was going to involve building scientific theories.

Readers are going to see a lot of discussion in this book about what might be understood by "theory" or at least a "scientific theory," which is much more than just a hunch or a guess, but one fairly certain thing is that most, if not all, theory building involves modeling. By this is meant, in a fairly straightforward fashion, making or conceptualizing a system or setup that is working, that is functioning. A clock would be a good example of a model, but so is a vegetable, growing and flourishing in the ground—and this is important. A model involves an analogy, in the sense that people take this system with which they are familiar and apply it to the phenomenon they are trying to understand, seeing the phenomenon as part of or produced by the clock system or vegetable system—or, rather, the phenomenon as part of or produced by a system like a clock or a vegetable.

It was the genius of the Greeks to come up with a model that explained the motions of the stars—and of the sun and the moon and the wandering stars—and to put it all into context with respect to the earth, the home of humans and of all other known living things. It is not obvious that the earth is a sphere. In fact, the creation story of Genesis rather suggests that the earth is below and the heavens are above. Indeed, apparently borrowing from Babylonian cosmology, various passages in Hebrew scriptures portray the earth as a flat expanse of land and sea covered by a vaulted heaven containing the stars and other celestial objects, none too high. But the idea that everyone before Christopher Columbus thought that the earth was flat is a nineteenth-century fairy tale, and certainly the ancient Greeks had growing awareness that the earth home is a globe. Apart from anything else, if people look out to sea, they see the waters drop off at the horizon—but if they take a ship and go out to sea, as many Greeks sailors did, they learn that the sea doesn't drop off but simply keeps going and the land behind starts to sink below the horizon. Only travel across a globe would give this illusion. Also there were eclipses of the moon. The obvious explanation of them was that the sun got behind the earth, which cast its shadow on the surface of the moon—and since this shadow was circular, it followed that the earth itself must be a globe. Most telling of all, the noontime shadows cast by obelisks or other tall structures at any given date and time would be longer farther north, and Greek geometers used this angle and the distance between the objects to compute with great accuracy the earth's circumference.

If someone starts with the idea that the earth is a globe or a sphere and it is obviously not moving, then it naturally follows that it is plonked right there in the middle of things. The stars are moving around and around, fixed relative to each other, so the equally obvious assumption is that they too lie on the surface of a sphere—a very large sphere that defines the outer bounds of space, and that sphere is circulating around a point, namely the earth, which is at the middle of things. Because the sun rather blots out or brightens things, people only see the stars at night, but everyone can readily imagine that they are there during the daytime—it is just that sunlight masks them. Now at this moment, a rather interesting question arises. The stars go around and around on their sphere. And they do this pretty much once every day. But do they go uniformly? What's to stop the stars from going faster during the daytime and slower during the night? To answer this question, it first has to be recognized that the Greeks didn't separate out what people now call "science" from what they call "philosophy" or might call "religion." As good if not better at mathematics—geometry particularly—than just about anything else, the Greeks had fixed on the circle as something special. They thought it was perfect, and accordingly they thought the perfection demanded a steady state of rest or of motion. If the stars sped up or slowed down, they wouldn't have the perfection that is desirable. Desirable to whom? Well, that is a good question—at least in some sense desirable to the ideal philosopher, whoever or whatever that might be. More on this point in a moment.

THE SUN'S PLACE

So here is the basic picture or model of the ancient Greeks: the spherical earth fixed at the center, and the stars revolving uniformly on an outer sphere. Now things start to get more interesting and more complex. First, what about the sun? At one level, the answer is easy: the sun rises in the morning as the stars start to vanish, goes through the heavens, and then sets in the evening, when the stars start to come out again. Why not simply put the sun on the sphere and let it spin around like the stars? Well, yes, but during the year, the sun rises and sets at different times, and likewise during the year sometimes it is higher in the heavens at noon and sometimes lower. And with respect to the constellations, sometimes (as judged by the location of the stars when the sun rises or sets) it seems to be in one part of the heavens and sometimes in another part. In short, to observers on earth, even as the sun circles the sky each day east to west, it appears to move steadily and dependably backward (or west to east) among the fixed stars about 1 degree per day, returning to its original location among them over the course of each year.

The genius breakthrough came when some Greek natural philosophers realized that this could all be explained by putting the sun on its own sphere nested within the stellar sphere (or firmament) and by having the sun move slightly slower (with respect to the forward motion of the stars), a full cycle taking one year. This meant that the sun's motion is mostly forward with the stars (the "diurnal" motion), but slightly backward in its own right along the ecliptic (the path the sun apparently takes along the surface of the sphere of the stars). Since the length of the day is fixed by the sun, and since the sun's backward motion adds on four minutes a day, this means that the diurnal motion of the stars takes twenty-three hours and fifty-six minutes to complete a revolution. One more tweak is needed to get the effect of the sun's changes during the year. Instead of the ecliptic being simply on the equator of the outer sphere, as it were, it is slanted so that during the year the sun takes different positions with respect to the heavens and takes different arcs during the day with respect to an observer always looking upward from the same place on earth. Notice that optical illusions come into play here.

The rub, if they viewed it as one, came in trying to envision how this mathematical two-sphere model might work physically. No one really wanted the sun embedded in the outer stellar sphere, plowing its way ever so slowly backward through the home of the stars. That would not make the firmament very firm at all. But suppose the sun were embedded in its own sphere, within but concentric with the outer stellar sphere, with both spheres made of some kind of hard crystalline substance that people cannot see, like two nested glass globes, one with stars on it and one with the sun on it. Then once set spinning, and because of the distance and the transparency of the spheres, the effect to earthly observers would be as if the sun were actually moving backward among the stars. Note also that the demand for uniform circular motion fits in nicely. Although the sun's motion is now a compound, it is a compound of two uniform circular motions. Its sphere spins slowly from west to east, but then on top of this, it is carried along with the outer stellar sphere spinning much faster east to west. Or perhaps there could be two spheres, both spinning one way but at different speeds. Either physical model accounts for or "saves" the observed phenomenon.

ADDING TO THE CELESTIAL PICTURE

This is just the start. Next up is the moon, and after that the moving "stars" that later were recognized as planets within the solar system. The way forward is pretty clear: give them all their own spheres and set them spinning, too! What about the looping backward or "retrograde" motion of these planets? The ingenious, mathematically gifted Greeks discovered that they could account for this optical effect if they added enough spheres—some nested within and others (called "epicycles") riding along the edge of other spheres like little ball bearings—and set them spinning in different ways, all with uniform circular motion. So they end up with something a little bit like a perfectly round onion, with layers of concentric crystalline spheres festooned with epicycles, all circling merrily away but in a steady sort of state. When the medieval Catholic Church latched on to this model, some of its theologians saw a role for angels, archangels, and other spiritual beings from the Bible, pushing these spheres as they joyously contemplated the divine presence within and beyond. Scientific ideas thus fostered religious beliefs and served as ready proof for the existence of God. Surely some supreme being had to design and maintain such a complex and beautiful celestial structure.

Note that this picture or model provided an immediate answer to something that is very striking. Down here on earth nothing stays very much the same. There are the seasons, not to mention natural calamities like storms and earthquakes. Up in the heavens everything appears very much the same. Beyond their fixed rotations, the stars don't change, and neither does the sun or the moon or the wanderers. Their positions change, obviously, but their natures seem—well, they seem to be eternal. The model explains this easily. Up beyond the moon everything is unchanged and eternal. The crystalline spheres just keep moving along, and all is as it was before and as it will be in the future. The movement of the moon however stirs things up down here—the tides are something obviously tied in to the moon and its motions—and so nothing is stable. The crystalline spheres obviously end at the moon. The earth is not embedded in anything. It rests at the center of everything. So Greek natural philosophy had a solution to something that might otherwise seem puzzling.

Of course, there are questions left dangling. The ancient

Greeks were not so much interested in origins. This is more a Jewish and, later, Christian and Muslim way of thinking. For the Greeks, the great ideal was mathematics, and no one asked when 2 + 2 = 4 started to become true. The concept is outside time and space and is eternal—not everlasting in the sense of very old, but eternal in the sense of never old or young for that matter. Greek natural philosophers tended to think of the universe in this eternal way. It is true that Plato argued that there is a god, the Demiurge, who ordered everything, but there is reason to think that for him this was less a physical entity and more a principle of ordering. In the same way, someone might say that God made 2 + 2 = 4 without implying that one day in the past God made a choice between 2 + 2 = 4 and 2 + 2 = 5. God stands behind mathematics as the Demiurge stands behind the universe.

But origins apart, there is still the question of why the spheres keep spinning. Why don't they slow down and stop? What keeps them in business, as it were? It is clear that the Greeks thought some kind of divine agency was at work here, but how and what was another matter. Sophisticated Greeks had moved beyond the Olympus gods, Zeus and that sort of character. But they still looked to divine agency. For Plato, standing behind the Demiurge, perhaps in ways identical to it, is what he called the "Form of the Good"—a kind of spirit force from which all else stems. He was fond of analogies with the sun, arguing that just as the sun gives light and life in the physical world, the Form of the Good gives light and life in what Plato thought was the world of true being, the world of rationality. Aristotle, Plato's student, had something similar, although a little more complex. But for both them and others, it was a kind of divine force that made things work.

More troublesome was the matter of getting quality and quantity in harmony. Even accepting all of these different spheres and adjusting them to give effects like retrogression, it was still difficult to fine tune things to get the exact observed measurements of the planets (never mind the physical implausibility of the system). At what speeds must these spheres spin in order to get the precise positions that are observed of the heavenly bodies? Here things started to get very complicated—so complicated, in fact, that those who went at these matters professionally often turned to another model for help. Instead of getting effects like retrogression from concentrically spinning spheres, some Greeks turned to what mathematicians call "deferents" and "epicycles." They set a point going around an earth-centered circle—the deferent. Then around this point they set another point or planet cycling on its own path—the epicycle. They then got the kind of effect that one has when barbed wire is unraveled—looping the loop over and over again, overall forward motion, but at any point the planet might as well be going backward as forward. Using this picture, things were much easier to tackle mathematically, especially by adding epicycles on epicycles—an epicycle at one level serving as a deferent at another level—plus, as some Greeks did, accepting an off-center center of uniform angular motion for some of these deferents, called an "eccentric." Once they began tinkering with the model to better account for the observed motion of the planets, why stop until perfection was reached?

The trouble then, of course, was that they were in an even greater quandary when they thought causally. How could deferents, epicycles, and eccentrics work, since they seemed to blow apart any hope of bodies embedded in nested crystalline spheres? The epicycles cut right across them and the eccentrics upset the very notion of physical nesting. So there was a tendency to adopt what philosophers of science (especially and understandably those in the pragmatist tradition) call an "instrumentalist" approach to things. Put observations in and get predictions out, and don't worry too much about what goes on inside. This approach does not focus on the truth or falsity of theories or models, but more on their pragmatic or instrumental value. Modern readers might think that this is all rather unsatisfactory, but do note that this goes on all of the time in science. For instance, nobody thinks there can really be square roots of negative numbers, but in certain branches of science—electronics for instance—work would virtually grind to a halt if they were ignored.

THE METAPHORICAL NATURE OF SCIENCE

One of the most interesting things about science—and remember it is the philosopher between us taking the lead here—something that the Greeks (especially Aristotle) caught on to is the extent to which it is metaphorical. We hinted at this phenomenon when talking of models as analogical. This is very much part of a broader picture of seeing scientific understanding as taking ideas from one domain and applying them to another. For example, strictly speaking, when people talk today of a magnet "attracting" iron filings, no one or thing is doing anything of the kind. Romeo was attracted to Juliet, Napoleon was attracted to power, and most people are attracted to ice cream. Attraction involves someone being drawn emotionally to someone else or something. Only in the metaphorical sense does this describe magnetic attraction. Certainly iron filings appear to have a compulsion to get as close as possible to the magnet: they move across the space in the direction of the magnet and are hard to remove once there. In an earlier age, before the rise of the mechanical philosophy in the 1600s, natural philosophers like William Gilbert may have envisioned magnets attracting iron from internal forces, much like the notion of romantic attraction, but that went out with René Descartes. All that remains is the metaphor.

Students of metaphor may point out that not all metaphors are equal. In particular, some are more encompassing than others—the cover for many others, or what are known as "root" metaphors. Consider how goodness and health are thought of in terms of standing erect: "He was an upright citizen." "Stand up, stand up, for Jesus." And the Shriners' motto: "A man never stands as tall as when he kneels to help a child."

What's the root metaphor that the Greeks were working with to explain the universe? Since the sixteenth century people might equate it to an old-fashioned mechanical clock with all of those gearlike spheres going around and around perpetually driving the whole. To the modern mind, this is the paradigmatic example of a machine. The fly in this ointment is that the Greeks didn't have clocks! Sure, they had some machines, especially for warfare, but overall theirs was not a mechanical society. There was, however, another candidate for understanding the physical universe, and that was the organism. Take Mother Earth herself—*Mother* Earth! Go back to the early Greek farmers at the beginning of this chapter. Along with the passage of the hours, they also had the passage of the days and weeks and months. Earth does not remain constant, even in warm climates like Greece. It has seasons of growth, flourishing, harvesting, decay, and closing down—and then it starts all over again. It has times when there is flooding and times when there is drought. It has times when the sun seems never to set and, in some parts of the world at least, when it seems never to rise. It has springs bubbling up out of the ground; it has rivers and lakes and marshes and fens, and the sea and its motions and rhythms. From this, it was not hard to see the earth as a living organism. And once the Greek philosophers went that far, why not extend the metaphor to the heavens? They are, after all, as much in motion as any other living thing.

FINAL CAUSES IN NATURE

Plato was quite explicit in thinking of the earth—the whole universe, indeed—as an organism. Aristotle was a little more circumspect, but certainly he thought of the physical as well as the biological world organically. And this leads to some interesting issues about causes. Consider Aristotle's thinking that causes are at the heart of explanations. They are the things that explain why events—effects—happen. The window is smashed. Why? Because some kid hit a ball through it. The moon goes dark during an eclipse. Why? Because the earth casts a shadow over it. Now notice something interesting about causes and effects. The cause always occurs before the effect, or at most at the same time. The bat was swung, the ball was hit, and the window was smashed. The earth passed between the sun and the moon, casting a shadow. And so on. The cause never comes after the effect—and there is good reason for this. If the window smashed and then the kid swung and missed, there is no explanation for the broken window. If the moon went dark but does not usually shine by reflecting light from the sun, there is no explanation for its darkness. The cause must come first; the effect follows.

But do they? Not inevitably. Think about organisms and what is fundamental about all of them: they reproduce, or at least they *can* reproduce, and that distinction is crucial. Some organisms have reproductive organs that are not reproducing most of the time—so why do they have them all of the time? What are the causes? The answer is that living organisms have them in order to reproduce at some point in their lives. Reproductive organs are thus explained by Aristotle in terms of events that could happen in the future, events that may never happen but in some natural sense should happen.

Aristotle called these sorts of causes "final causes," and they are not incompatible with—rather, are complementary to—regular causes, or what Aristotle called "efficient causes." In modern terms, the efficient causes of human reproductive organs are genes. The final causes are newly produced human beings. Note also that the problem of causes not occurring—what is sometimes known as "the problem of the missing goal object"—is not the devastating objection that comes with efficient causes. If the kid does not hit the ball, then there is no cause for the broken window. However if there are no babies, and that is still what the reproductive organs were for, it is just that they never produced a child. Now Aristotle maintained that each thing had but one final cause, and, following him, the late medieval Catholic theologian Thomas Aquinas concluded that reproduction was the sole purpose of sex and any steps taken to frustrate that end would be sinful. Protestant reformers rejected such thinking and accepted other purposes for sex within marriage, including pleasure. Immanuel Kant had the right idea when he said that final-cause talking is usually appropriate when there is a cause-and-effect chain—A leads to B leads to A and so on—in our example, the gene leads to the new human leads to the gene and so on—but not otherwise or in exclusion of other causes.

One last point. Thus far we have been talking about final causes in the organic world. The Greeks—Aristotle in particular—wanted to use them in the physical world also. The rain exists in order to flood the fields and thus make growth possible and so forth. His use of final causes comes through most evidently in his physics, based as it was on the old idea that all matter is one of or a combination of four elements—earth, water, air, and fire. Aristotle argued that everything has its proper place in the universe, with earth at the center, and then in order water, air, and fire. So when a stone or any heavy earthy object is released, it falls down in order to get to its proper place, namely the center of the universe. The smoke from a fire rises in order to get higher up in the scheme of things above more earthy things. In between, there is water on top of earth and air on top of water. Here a thoroughly organic metaphor pervades Aristotle's thinking. It points toward his thinking about God or gods. There is an Unmoved Mover (or in some versions Movers) that exists independently of the universe and that may indeed have no interest in or knowledge of the universe! However, this Mover is perfect, and because of this the universe strives to unite with it. It is the cause of everything, but not the efficient cause. It is rather the final cause. Aristotle did not need angels and archangels to spin the celestial spheres: spinning was simply in their nature, much like it was in the nature of iron to move toward magnets.

Moving on now, it goes almost without saying that although this new religion was totally unknown to the ancients, when Christianity arrived on the scene, the early philosopher-theologians like Augustine found much that was very congenial. Most particularly, the two-sphere model—which came to be known after one of its most distinguished practitioners as "Ptolemaic" astronomy—located the earth right at the center of all of the action, and it made the crucial distinction that down here on earth all is change and strife and decay due to the Fall, and up there in the heavens all is eternal and unchanging and perfect, as God intended. This was not the cosmos of the Jewish scriptures, but it became the cosmos of the medieval Catholic Church. Obviously the Christian God is neither the Form of the Good nor an Unmoved Mover indifferent to the universe, but here too there was much room for creative interaction. It was not as if the Christians had to make their God compatible with Zeus and his various actions and intentions, much of which involved intercourse with any female who happened to catch his eye. They could see their God as perfect, and see God's son and the Holy Spirit as perfect too—the triune co-creatures of the universe, with us humans perfect in inception but fallen on earth due to human sinfulness.

Islam came about somewhat later but within a culture shaped by Greek thought and a Ptolemaic cosmology, even more than Christianity was. Late Roman and medieval Catholics needed to baptize Ptolemaic cosmology; they could not find it in the Bible. In contrast, a Ptolemaic worldview runs through the Qur'an and is still taught in some traditional Islamic schools. The earth is central and heavens circle about in regular, predictable, and infinitely perfectible patterns.

THE HELIOCENTRIC THEORY

It is therefore no surprise that for the first fifteen hundred years of its existence, the Christian religion fit happily with the cosmological legacy of the Greeks (except for Aristotle's commitment to the eternity of the universe). To a certain extent, Orthodox Christians still do fit happily, given their reluctance to relinquish the Julian calendar, which is based on the Ancient Greek worldview. Why then did things change? And why in particular did the Polish astronomer Nicolaus Copernicus publish a book in 1543 that overturned this picture, putting the sun at the center of the universe and displacing Earth as one of the planets—the others being the wandering stars—circulating around this central body? Two possible reasons why there was a rejection of a "geocentric" theory of the universe for a "heliocentric" theory can be easily dismissed. First, it was simply not the case that a whole new pile of empirical evidence came along refuting earlier thinking and making modern thinking much more plausible. By the sixteenth century, Ptolemaic astronomical predictions were out of whack with the observed motions of the planets: what had seemed minor discrepancies a thousand years earlier were now glaring anomalies due to the accumulation of small deviations over time. Ptolemy could be patched up by fixes, but no one had the mathematical skills to do so until Copernicus came along, and he chose to use a different structure rather than fix the old one. Second, it was simply not the case that the eruptions in religion, in particular the Protestant Reformation, had much to do with anything in cosmology, either. Although not an ordained priest, Copernicus was a lay Catholic official who lived and died within the church. No one was becoming an atheist or even an agnostic and throwing over the central status of human beings. Protestants were if anything even keener than Catholics on the old order of things; Orthodox Christians even more so. It was not even that no one had hitherto thought of the heliocentric theory. A couple of centuries after Aristotle, Aristarchus of Samos had floated just such a theory—one that got little or no traction at the time—possibly picking it up from earlier Pythagorean thinking.

A major factor in Copernicus's choice was as much aesthetic, philosophical, or theological as anything. Back in antiquity there were always people who wanted to elevate the sun to something more than just another object in the universe. The Pythagoreans in particular were, if not outright sun worshipers, a group that gave the sun very high status indeed. This was no sillier than, say, considering the earth to be an organism. Further, the sun provides light so that people can see. With reason, many thought this to be the highest of all of the senses. On the other hand, the sun provides heat and fuel and much more, making plants and animals grow, and generally being the life source of the world within which people live. Without the sun—as people realized in particularly rainy years—nothing flourishes and grows, and people suffer and starve and die. Importantly, Plato picked up on all of this, and his dialogue *The Republic* particularly makes much sense when read in a Pythagorean light, from the veneration of mathematics to placing the sun in the highest position here on earth, the counterpart to the Form of the Good in the rational world.

With the coming of the Renaissance there was a return to the original Greek sources, rediscovery of ancient tracts, and a corresponding rise in the status of Plato. For someone like Copernicus, who was already deeply initiated into the mysteries of mathematics, a Pythagorean-Platonic empathy was almost inevitable, and with this came a willingness to elevate the sun to the highest position in the universe. Not that this exhausted all that there was to be said for the heliocentric system. Most particularly, in one fell swoop, Copernicus solved one of the outstanding problems of the older position—the reason for the distinction between the inferior and the superior planets. Two of the planets, Mercury and Venus, seem never to stray far from the sun, and when they retrogress they do by moving back across the face of the sun. Three of the planets, Mars, Jupiter, and Saturn, move all over the heavens and only retrogress when they are far from (in opposition to) the sun. For Copernicus, the reason for the division fell out immediately. Mercury and Venus, the inferior planets, are closer to the sun than Earth is, and Mars, Jupiter, and Saturn, the superior planets, are farther from the sun than Earth is.

Moreover Copernicus confirmed what had been known but little discussed by astronomers all along. When planets retrogress they appear brighter to observers on earth—and not just a little brighter, but especially for Venus and Mars, much brighter. With crystal spheres and a geocentric universe this shouldn't happen, because heavenly bodies are unchanging and any individual planet would always be the same distance from Earth. With deferents and epicycles one has a rough explanation—planets are closer to Earth, and hence brighter, when they are retrogressing than at other times, but the planets actually appear much brighter than even epicycles would allow. In a heliocentric system, with Earth and other planets moving around the sun, the vast changes in distances between Earth and other planets following their separate paths around the sun would readily (and fully) account for the apparent changes in brightness. Copernicus drove in this stake more firmly because on his system the planets are indeed closer to Earth when retrogressing than at other times. Observation fit theory; that is what scientists like. Theologians might not be so happy if the old theory was wedded to religious beliefs—though it would ultimately prove helpful that those beliefs were not integrated into the Jewish and Christian scriptures, which was not the case for the Qur'an.

RECEPTION OF HELIOCENTRISM

Philosophy is one thing. Convincing people is another. On this, even the philosopher between us agrees and the historian is adamant. The ancient Greek Pythagoreans were somewhat of a philosophical or quasi-religious cult, priding themselves on having esoteric knowledge that was grasped only by them. Since they were heavily into mathematics, they didn't have too much trouble on this score. Plato was obviously influenced in this direction, because in *The Republic* he not only makes much of mathematics but stresses that those who will have mastered the subject—the Guardians—and who are now going on to learn about the Forms and the role of the Form of the Good will likewise be possessors of knowledge not generally available. There is reason to think that this was Copernicus's attitude too. Certainly he made little effort to make his thinking user-friendly in his grand treatise, *De revolutionibus orbium coelestium*, which was published just before his death in 1543. What does seem clear is that there was no general rush to his position. One historian calculated that in sixteenth-century Europe only ten people became Copernicans. Interestingly, five of these were Protestant and five Catholic. None were Eastern Orthodox or Muslim, even though many scholars of each faith then lived in Europe.

To many, it certainly was not obvious, a priori, that acceptance of heliocentrism was mandated. At the psychological level probably the biggest barrier was the huge extension in the size of the universe demanded by Copernicus. If Earth is moving rather than stationary, then people should see some changing difference in the positions of the stars depending on where Earth is in its orbit around the sun at various times. It is rather like looking at an object in a room from different positions and different angles. If observers don't see any difference, then either this is because there is no difference—Earth is not moving—or they are so far from the stars that they cannot perceive if Earth has moved except with highly refined telescopes, which did not then exist. Without telescopes, astronomers could see no differences, and so they calculated that the universe must be at least four hundred thousand times bigger than earlier estimates that had observed celestial movement with the naked eye. Since the Greeks thought in vast terms already, imagine what a mouthful the new implications were to swallow.

DEVELOPMENTS AFTER COPERNICUS

More important than the rush or non-rush to heliocentric thinking were the developments in the years after Copernicus published in 1543. First there was the late-sixteenth-century Danish astronomer Tycho Brahe, who mapped the heavens with great accuracy—incredible, really, for not having a telescope. No longer could there be excuses about the flabbiness of the available data. Then his assistant, the German-born Johannes Kepler, broke with the tradition of insisting that heavenly motions must be circular. Famously, he found that the planets circulate the sun in ellipses, with the sun at one focus. Finally, on top of all of this, the early-seventeenth-century Italian astronomer Galileo Galilei invented—more probably refined—the telescope and discovered wonders like the moons of Jupiter, phases of Venus, and the terrain of the moon. If his observation of Venus having phases was accepted—and not everyone did at first—then at least Venus (and, following Galileo's logic, by implication Earth and other planets) must go around the sun. Galileo also developed modern mechanics, showing how matter in motion obeys certain laws and not others, with these new laws allowing for Earth to orbit the sun without anyone on earth necessarily noticing it. Most importantly, and that which got him into trouble with the church, he wrote in the vernacular, popularizing the Copernican system. As is well known, he was condemned and put under house arrest. This should not have happened—and for all that, there have been apologies, and it has been an embarrassment to the church ever since—although general opinion among scholars is that Galileo was not always an easy person to get along with and in major respects he courted official disapproval of his ideas.

What Galileo as a scientist was doing was breaking down the traditional distinction between the perfect unchanging heavens and the imperfect changing world in which people live. This paved the way for the seminal contributions of the English mathematician Isaac Newton, who in the second half of the seventeenth century brought what has become known as the Scientific Revolution to a triumphant conclusion, when with his law of gravitational attraction he showed how everything—down here on earth and up there in the heavens—can be explained by one overarching mathematical hypothesis. Bodies attract each other by a force inversely proportional to the square of the distance between them. With that great insight, the heliocentric picture of the universe was given a firm mathematical basis.

A NEW ROOT METAPHOR

Let us pause for a moment and ask about some of the implications of all of this. First, the philosopher among us asks, What was going on at the more theoretical or metaphorical level? The old two-sphere system, the Ptolemaic system, had the organism as its root metaphor. This made talk of final causes natural and acceptable. The new breed of scientifically minded natural philosophers wanted none of this. Final causes in particular came in for scornful treatment, with the philosopher of the Scientific Revolution, Francis Bacon, likening them to vestal virgins—beautiful but barren. The basic problem was that no one could see much use for them, so they were being pushed out of science—or at least out of physical science. (As will later be discussed—especially in chapter 9—the story was somewhat different in the biological sciences.) But if final causes were going, what then of the organic root metaphor? At this fundamental level—perhaps the most fundamental of all in the Scientific Revolution—the root metaphor changed from the organism toward the machine. What was happening was that people were no longer thinking of the world in terms of vegetables or animals but beginning to think of it in terms of contrivances, of human-made systems designed to perform certain functions perpetually.

This was now the age of the mechanical clock in Western Christendom, and the theoreticians and philosophers of the Scientific Revolution seized on it. The seventeenth-century Anglo-Irish aristocrat Robert Boyle, distinguished as one of the leading figures in the history of chemistry, was scathing about those who relied on final causes and hence somehow were thinking in terms of (what he would have regarded as occult) forces driving nature. In "A Disquisition about the Final Causes of Natural Things," published in 1688, he said, making reference to one of those wonderful church timepieces, "The world is like a rare clock," he said, "such as may be that at Strasbourg, where all things are so skillfully contrived that the engine being once set a-moving, all things proceed according to the artificer's first design, and the motions of the little statues that at such hours perform these or those motions do not require (like those of puppets) the peculiar interposing of the artificer or any intelligent agent employed by him, but perform their functions on particular occasions by virtue of the general and primitive contrivance of the whole engine." By this, Boyle did not intend to push God back solely to the beginning, a mere clockmaker, but as a devout Protestant he rather liked getting rid of needing Catholic angels and archangels to push around the heavenly spheres. Indeed, none of the major figures in the astronomical revolution—Copernicus, Brahe, Kepler, Galileo, Newton, or Boyle—were atheists, agnostics, or deists. They all considered themselves as sincere, right-thinking Christians, though Newton was a rather unorthodox one.

GOD'S ROLE IN NATURE

The physical universe is like a machine not an organism, early modern natural philosophers suggested. They had entered the mechanical age. Move now from philosophy to religion. What then of God? Well, obviously the Copernican Revolution made the universe a lonelier place for humans. It suddenly became absolutely huge—it wasn't long before people started to think infinite—and Earth was no longer at its center. It was just one planet among several circulating the sun and—some soon speculated—perhaps the stars were also suns, with their planetary systems, and some of those planets might have life. Obviously this didn't disprove the existence of God, yet God's concern for humans and their special status in the universe became less obvious. But surely what was lost on the one hand was gained on the other. Although the earth seemed a bit downgraded, the general metaphor was, if anything, even more God-friendly than before. Machines have machine makers, in this case God, and inasmuch as the universe was seen to be that much bigger and more magnificent, so God was seen to be bigger and more magnificent, not physically, of course, but in intention and scope and ability. It was not atheism or deism that drove the Scientific Revolution. Virtually all of the key figures in the Scientific Revolution were Catholic or Protestant Christians who saw their work as glorifying God and defending the faith. This is not to say Christianity necessarily leads to science, of course, since Eastern Christians did not participate in making modern science. For his part, and it was a central one, Newton even denied the clock metaphor for the cosmos insofar as it implied that once constructed, the universe could continue without the ongoing assistance of God. His God was real, omnipresent, and omnipotent.

Yes, but! The trouble is: once God was made into the Supreme Machine Maker, then as was made very clear in the passage from Robert Boyle quoted above, God increasingly was taken out of the day-to-day workings of the universe. He got it all up and running and then sat back and didn't interfere. For Boyle, God still did maintain a loving, active relationship with people, and perhaps even had more time for that sort of stuff now that the universe ran on its own. But before long, and despite Newton's protests to the contrary, increasingly God was being pushed out of science, and naturalistic explanations became the sole object of those doing science. There was a stampede to atheism, deism, or agnosticism but no further compulsion to keep thinking about spiritual (or even final) causation in nature. In the words of the historian of science D. J. Dijksterhuis, increasingly God became a "retired engineer."

ACTION AT A DISTANCE

In short, Newton conquered. His mechanics became the paradigm for what a physical science should be. And yet it was not always easy sailing, particularly at first. The followers of the early-seventeenth-century French philosopher and mathematician René Descartes—Newton's predecessor by a generation—insisted that true mechanism eliminates all hidden or occult forces. This means in particular that if one body acts on or in any way impinges on another, it must do so directly. Like gears in a mechanical clock, the moving parts must touch. There is no place for "action at a distance," that is, forces or causes working across spaces. And yet that idea is precisely at the heart of Newton's mathematical law of gravitational attraction. That law treats bodies as having effects on each other, without touching, across space. Newton did posit an explanation that provided for a mechanical connection between these bodies in the form of an intangible ether pervading space, but he deemed this just a hypothesis, not even a theory and far from a law. Gravity was true, Newton maintained, and he did not feign to know how it worked.

This was quite unacceptable to the Cartesians, and it was only as the eighteenth century went on that slowly the huge predictive powers of the Newtonian system made the theory nigh irresistible. Action at a distance was judged compatible with mechanism. Although, looking back, perhaps the Cartesians did have a point. When he was working on his science, a close companion of Newton was the philosopher Henry More. A leading member of what are now known as the "Cambridge Platonists," More had no trouble with occult forces and biological root metaphors. His enthusiasm may indeed have infected Newton and given him courage to push forward with his own thinking about the nature of gravitation. Newton was no Aristotelian, but perhaps the organic metaphor was not as firmly excluded from modem science as the enthusiastic mechanists supposed.

THE UNIVERSE AND BEYOND

In a way, that is the end of the story. In a way, that is just its beginning. It is the end because, Newton's action at a distance notwithstanding, the machine metaphor dominated in physics generally, and in astronomy and its causal discussions, cosmology, specifically. The world works according to unbroken law and, for most modern astronomers and physicists, God stays out of it. All is efficient cause. There is no place for final cause in modern astrophysics. Some physicists and astronomers see God as the Creator of those laws; many do not; but in either event, God is not part of doing modern astrophysics. It is the beginning of the story because, although for two centuries Newton's vision of the universe reigned supreme, by the dawn of the twentieth century, thanks above all to the German scientist Albert Einstein, it all came tumbling down. With the historian between us taking the lead, we will look more deeply into those matters in the next chapter, but a few closing remarks are appropriate here because they so closely relate to cosmology.

First there was Einstein's special relativity theory. The speed of light had been measured, and it was found that no matter the motion of the observer, it would always be the same. To explain this, time was included as a dimension along with the three dimensions of space, and some paradoxical (although later confirmed) conclusions emerged—most striking being that time for an individual on a spaceship moving really fast would go more slowly than for the twin left stationary back at home. Then, second, there was Einstein's general relativity theory, with its (seemingly) incredible assumption that space is not Euclidean but in some sense curved by gravitational fields, and the consequences thereof. On the basis of this theory, predictions could be made—predictions subtly different from those of Newton—and rapidly it found acceptance. Then quantum mechanics came along, which even threw Einstein for a loop. The certainty of Newtonian dynamics gave way to uncertainty at the smallest levels, which, of course, are the foundation for actions at every larger level. Einstein saw this as God playing dice with the universe, which he could not accept, but some deeply spiritual physicists—especially in

Eastern religious traditions, but also some Western theists—saw it as providing a way for God to act on physical matter in a causative fashion without being detected or detectable. This did not open a wormhole into science for God, but it could keep science from ruling out the divine in nature altogether. God could be part of the Heisenbergian uncertainty about whether a particle is a mass or a wave, the unseen actor within space and time, which suits some religious believers just fine.

One of the most interesting implications of the new physics was that it was now possible to get a firmer handle on the origins of the universe. Newton thought that every now and then things needed to be adjusted by the Divine Hand, but from a scientific viewpoint the universe could have been eternal—as Aristotle thought. In the eighteenth century, thanks perhaps to developments in geology, people turned their attentions more and more to origins, and it was not long before speculations started to appear that tried to trace the present state of the universe back to earlier times. Prominent was the so-called nebular hypothesis, proposed by a number of thinkers including the German philosopher Immanuel Kant and the French mechanist Pierre-Simon Laplace, supposing that the solar system was formed by the coalescing of a nebula, a great cloud of gas of a kind observable in the night skies. A version of this theory is still held today, extending it from the solar system to all such systems.

The nebular hypothesis does not explain the origin and existence of matter as such, however. In the twentieth century two rival hypotheses were proposed. One, the steady-state hypothesis, championed by the English astronomer Fred Hoyle, supposed that matter is continually being produced throughout the universe. The other, the Big Bang hypothesis, indebted to Einstein's thinking and first proposed in 1927 by the Belgian astronomer (and Catholic priest) Georges Lemaître, supposed that everything was first condensed down into one superdense point and then blew apart, ever expanding and forming the universe as people now know it. Both hypotheses were scientific in nature, not religious, but religious believers saw implications in them, with some deists and liberal theists favoring the steady-state view, while Catholics and mainline Protestants seeming to welcome the idea of a big bang at the starting point of creation. After much acrimonious discussion, gradually the physical evidence of low-level background radiation emanating from the edges of the expanding universe pointed unambiguously to the second hypothesis.

Astrophysicists now consider that the universe began about fourteen billion years ago in a colossal big bang that is still exploding outward, with ever mounting evidence confirming theoretical predictions based on that view. Science works in this matter: speculative hypotheses tested by observations and experiments leading to established theories. Not all Christians welcome this theory, however, with some Protestant fundamentalists abhorring its supposedly non-biblical chronology. Less literal Christians, however, can see the Big Bang theory as rough affirmation of the first verse of Genesis: "In the beginning God created the heavens and the earth. . . . And God saw that it was good"—*Bang!* And of John's Gospel: "In the beginning was the Word, and the Word was with God, and the Word was God"—*Bang!* Simply having science put a beginning on the physical universe affirms or at least leaves open the possibility of a creation, and to many, a creation implies a Creator. Sustaining such a belief is central to the more-existential, Kierkegaardian forms of the religions that hold the Genesis account sacred—Judaism, Christianity, and (by reference via the Qur'an) Islam—and may even represent the ultimate leap of faith that gives those religions meaning. For if God did not create the physical universe, or at least the laws of physics that led to the universe, then how can anyone believe anything or at least that anything is good, some theologians argue. And if God did create it or them, then how can anyone question its or their goodness even in the face of seeming evil? Pottery cannot question the potter, scripture tells Christians and Jews, even in the most challenging situation.

But in all this it is important to remember that the Big Bang is a scientific theory. As such, it cannot prove or disprove religious claims. In particular, it says nothing certain about the creation story in Genesis. Apart from the fact that they are completely different stories, their intent is quite different. The Big Bang theory explains the physical nature of the universe. Genesis and like stories address the metaphysical question of why there is anything at all. Why is there something rather than nothing? People accept the Big Bang theory and still wonder why there was the original compressed pinpoint of matter in the first place. And if they give a scientific answer—for instance, if they say that after the Big Bang the universe expands until it runs out of steam and then starts to contract into the Big Crunch, at which point it all starts again—that does not answer the metaphysical question. Why is there Big Bang, Big Crunch, Big Bang, Big Crunch . . . ? This remains a scientific question—and one that may be beyond the realm of science to answer because it would involve peering back before the beginning of time. Yet it too has religious implications. Eastern religions in particular appreciate the idea that the universe may be cycling between existence and non-existence, or be periodically reincarnated.

As it happens, there seem to be several possible theoretical answers on the table at the moment. Perhaps there is the ongoing cycle, as sketched in the last paragraph. Perhaps there was—and for all anyone knows still is—a state where more than one universe is created in an ongoing way. If this latter is true, then there is the exciting—exciting for some, at least—thought that there may right now be more than one universe, perhaps very many (infinitely many) indeed. This concept of the "multiverse" is highly controversial in science, especially if we suppose that different multiverses might have different laws of nature. Scientists line up on both sides of the divide, with philosophical, metaphysical, and even religious arguments in play, as well as scientific ones. Indeed, how much can science even say about multiverses, given that there is no known empirical way to test their existence?

Our philosopher of science has led the way thus far, but this is perhaps a good note on which to end our survey of this topic, since we are leaving the domain of history and entering into speculation as to where science and religion may venture in the future. Other topics bring these issues down to earth, where historians dwell, and thus we turn matters over to our historian of science.

CHAPTER TWO

The Tao of Physics and Other Big Ideas

A CHAPTER on the history of astronomy logically leads into one on the history of physics because over time astronomy merged into physics. Where early astronomers mostly plotted the course of planets amid the fixed stars, cast horoscopes, and made calendars, later astronomers increasingly studied the physics of the universe. During the seventeenth century, Kepler discovered his three laws of planetary motion, Galileo revealed his concept of celestial inertia, and Newton announced his law of universal gravitation. The heavens were no longer seen as some perfect abode for God and angels but a material place subject to the same physical laws as on the earth. Yet while this tended to bring astronomy down to earth, it lifted physics up to the heavens.

Whereas astronomy once asked the big questions and seemed to hold big answers—questions and answers that inevitably had religious implications—the next focus became physics. What is the ultimate structure of the universe? How does it work? When did it begin? Where is it going? Why is there something rather than nothing? And is there room for God anywhere in this vast temporal structure that may include multiple universes? Except perhaps for this last questions—and some scientists dispute even that limitation—many now put these questions in the domain of physics. The answers have become so abstruse, however, that fewer and fewer people understand them, which can lessen the immediate sense of interaction with popular religion. While many Christians and Muslims can and do get quite upset when they hear biologists tell them that they descended from apes rather than from two fully formed humans named Adam and Eve, who were specially created in the image of God (more about that in chapter 6), they are more likely to ignore issues raised by physicists about string theory, dark matter, quarks, black holes, and antimatter, even though, if understood, these concepts might have much to say about religion. We won't try to speculate about future interactions on this front but, for purposes of this chapter written from a historian's perspective, we will instead stick to the past relationship. That should provide more than enough to discuss.

PHYSICS AND THE ENLIGHTENMENT

Chapter 1, on astronomy, discusses Newton at some length, and he offers as good a starting point as any for us to pick up the story in this chapter on physics, because he laid the foundations of classical mechanics by articulating his three laws of motion in *Philosophiae Naturalis Principia Mathematica*, better known simply as the *Principia*, published in 1687. The first of those laws set the tone for all: unless acted upon by an external force, an object will remain at rest or continue to move at a constant velocity. With this, physical objects became material things devoid of the spirits or souls that had seemed to indwell in them during medieval times, and nature became nothing more than the sum of its parts. Matter in motion became everything, or so some scientists claimed. Reductionism, or the conception that complex phenomena can be fully explained in terms of their less complex component parts, ruled. Add to that Newton's law of universal gravitation, which showed in mathematical terms how the attraction of material bodies to one another could equally account for the fall of an apple on earth as for the orbits of planets in space, and religious skeptics could see a world without gods. Living in Newton's shadow, German philosopher Immanuel Kant famously wrote in his 1784 essay "What Is Enlightenment?": "If it is now asked whether we at present live in an enlightened age, the answer is: No, but we do live in an age of enlightenment."

Enlightenment means different things to different people, and it certainly did not point toward atheism for Newton. To him, it purified Christianity of postbiblical heresies (including the Trinity) and established a role for God in nature not only as the master craftsperson who created and periodically adjusted the clockwork universe, but as the active supernatural agent (or source for the spiritual agency) that caused gravitational attraction. After all, he reasoned, matter cannot act at a distance in the manner suggested by his law of universal gravitation. There must be an intervening cause underlying but invisible to this purely mathematical law, and for Newton that cause was the God of the Bible as he read that work in his own non-Trinitarian way, which held that Christ the Son was created by, and therefore not co-equal with, God the Father. To hedge his bets on gravity's cause, however, Newton later hypothesized a subtle but material ether in space as an alternative account for gravitational attraction. In the end, reflecting his religious interests he devoted more time to (or at least wrote more words on) theology than science and, true to his convictions, refused Anglican Communion on his deathbed.

Others saw things differently, however. A thoroughgoing acceptance of Newton's law of universal gravitation led to a rich strain of scientific research and philosophical thought initiated by German polymath Gottfried Leibniz in the late 1600s and extending over two centuries. To account for gravitational attraction and other properties of matter, it posited *vis viva* as an indwelling force animating material bodies. For Leibniz, as for Newton, this reflected a religious commitment: in Leibniz's case, to faith in an all-knowing God who could and would create a perfect, self-sustaining universe. Kant became part of this tradition, which also led to extensive experimental research measuring *vis viva* in elastic and inelastic collisions and drove Leonhard Euler's development of the calculus of variations as a means to describe it.

Then there were the agnostics and atheists inspired by Newtonian physics. During the height of the Enlightenment in France, for example, physicist Jean d'Alembert and philosopher Denis Diderot, in their influential *Encyclopédie*, presented the history of European thought as a flowering of rational thought and unshackling from religious superstition culminating in Newton's science. French materialists Baron d'Holbach and Julien Offray de La Mettrie echoed this theme. "Disdain for the humane sciences was one of the first characteristics of Christianity," the Marquis de Condorcet complained in his 1794 *Sketch for a Historical Picture of the Progress of the Human Mind*, setting the tone for a perception of persistent and ongoing warfare between science and religion. Yet, while reflected to an extent outside France in the writings of Kant and English philosopher David Hume, such a view of science versus religion was not representative of Enlightenment thought throughout Europe.

Not only did Newton, Leibniz, and most other Enlightenment scientists reject atheism, Kant maintained that while reason cannot prove God's existence, he must nevertheless believe in God as part of his moral system. "Thus," Kant wrote in the preface to his *Critique of Pure Reason* in 1781, "I had to deny knowledge in order to make room for faith." Even in Hume, there are hints of deism toward the end of his posthumously published *Dialogues Concerning Natural Religion*.

Perhaps Kant best expressed the Enlightenment view of religion when he wrote in that preface to his *Critique of Pure Reason*, "Our age is, in especial degree, the age of criticism and to criticism everything must submit. Religion through its sanctity, and law-giving through its majesty, may seek to exempt themselves from it. But they awaken just suspicion." This approach suggests that it was the revelation-based authority of religion, especially the law-giving and regime-establishing Catholic Church, that aroused much of the opposition to Christianity during the Enlightenment. "As things are at present, we still have a long way to go before men as a whole can be in a position (or can ever be put into a position) of using their own understanding confidently and well in religious matters," Kant wrote with a guarded sense of optimism about his age. "But we do have distinct indications that the way is now being cleared for them to work freely in this direction, and that the obstacles to universal enlightenment, to man's emergence from his self-incurred immaturity, are gradually becoming fewer." The caution implicit in this comment was merited. Whatever Kant might think or hope about his age, rational, critical "enlightenment" remained a minority position. In fact, the eighteenth century was also an age of intense religious ferment, with a Pietistic revival in Germany, the Wesleyan movement in Britain, and the evangelical Great Awakening in English North America.

THE THEOLOGY OF ELECTRICITY

The self-taught genius Benjamin Franklin, America's first great physicist as well its leading diplomat, printer, inventor, and popular philosopher, exemplifies one Enlightenment-era view of the relationship between science and religion. At the time, electricity was a novel field, with systematic experimentation driven by new devices to create and collect electrical charges on globes of iron and flasks of water. As those experiments only became commonplace late in his long life, Newton never closely studied electricity, though, like magnetism (which he did study), the ability of electrically charged particles to attract or repel other electrically charged particles seemed somehow related to the attributes of gravity. Indeed,

Newton speculated that the electrical spirit, whatever that was, might be the ultimate cause of gravitational attraction.

The experiments themselves were sensational and did as much as anything at that time to popularize science among the upper classes. Rotating globes generated large amounts of static electricity, insulated wires or rods transmitted it across a room or stage, and newly invented Leyden jars collected and stored it. The result could be a mighty bolt that knocked down a line of soldiers or (in one case) monks. Alternatively, a charged rod could lift lightweight particles, while a charged boy or girl dangling from an insulating rope or standing on insulating wax could give an electric kiss. The mystifying and amusing options were endless, and the scientific questions were both genuine and readily studied by repeatable, testable experiments in laboratory conditions.

Importing electrical equipment to Philadelphia and systematically experimenting with it, Franklin became the first scientist to explain electricity as an ever present subtle fluid (somewhat like Newton's gravitational ether) that is contained within and flows between material objects. An excess of fluid creates a positive charge that repels other positive charges; a deficit of fluid creates a negative charge that repels other negative charges. Sparks fly when negatively and positively charged objects come in contact, yet Franklin saw this as natural because oppositely charged particles attract. A balance results when an object, like the earth or anything "grounded" to the earth, contains a normal amount of the fluid. His kite experiment proved that lightning could charge a Leyden jar and suggested that it could be drawn off through a grounded rod. Viewing the universal amount of this fluid as a constant that flowed among objects to do work, Franklin hit upon the idea of the conservation and transfer of energy that would feature prominently in physics ever after. For this, some historians consider him the most influential physicist of the eighteenth century. He certainly became one of the leading intellectuals of his day, lionized in Europe as well as America.

Instinctively, Franklin saw the natural balance and attractive properties of the electrical fluid as part of God's plan. "For, had this globe we live on as much of it in proportion as we can give to a globe of iron," he wrote in his letters on the electrical fluid, sent in 1750 to London and published there in the book *Experiments and Observations on Electricity*, "our air would continually be more and more clogged with foreign matter and grow unfit for respiration. This affords another occasion of adoring that wisdom which has made all things by weight and measure!" Franklin was not a Christian, however, and he wrote in his autobiography of rejecting the beliefs of his Calvinist upbringing in favor of Deism at young age: "There is one God who made all things," he quoted himself as concluding, and "he governs the World by his Providence." Although Franklin befriended evangelical revivalist George Whitefield during that minister's visits to Philadelphia during the Great Awakening, Franklin later quipped about Whitefield, "He us'd indeed sometimes to pray for my Conversion, but never had the Satisfaction that his Prayers were heard." To Franklin, revealed religion in general and the Bible in particular simply were not reasonable, and, as a child of the Enlightenment, he subjected everything to reason.

Also instinctively, Franklin's faith in science led him to

believe that (to use his words) "considerable beneficial uses" would flow from the study of electricity, even though beyond the lightning rod, which he invented, he could not foresee what they might be. For Franklin, research into the natural offered more promise for human betterment than meditation on the supernatural, and he regretted the shackles placed by organized religion on reason. Electricity, he might have said in anticipation of the advertising men, brings good things to life. In this prediction, the Sage of Philadelphia proved prophetic.

Living at the same time as Franklin, English revivalist and forerunner of Methodism John Wesley was as fascinated by novel experiments with electricity as anyone, but, showing another side of the relationship between science and religion, he drew his own distinctive conclusions from them. In his compendium volume *A Survey of the Wisdom of God in the Creation*, Wesley speculated that electricity was "the general instrument of all motion in the universe" and might "produce and sustain life throughout nature, as well in animals as in vegetables." In doing so, he depicted electricity as a second cause, with God being the first. Displaying an openness to new technologies that helped humanity, Wesley hailed Franklin's fire-preventing lightning rod and predicted that electricity would have medical uses. Throughout these writings, however, he stressed the religious value of electrical research. It showed the power and glory of God's creation, Wesley claimed, and exposed the limits of human reason. "How must these [experiments] confound those poor half-thinkers who will believe nothing but what they can comprehend. But who can comprehend how fire lives in water, and passes through it more freely than through air?" he asked. "It is all mystery, if haply by any means God may hide pride from men." God's ways are beyond reason, Wesley believed, and science—at least the type of descriptive science that Wesley liked—shows it. Faith alone can save.

Yet it was just such religious faith that Franklin sought to dispel with his researches into electricity, and he inspired the English materialist Joseph Priestley to follow suit. By the time Franklin met him in 1766, Priestley was known as a radical educator and proto-Unitarian minister with utopian views rooted in his faith in scientific progress and human improvement. Things were getting better, Priestley believed, and were bound to keep doing so because of the deterministic, Newtonian laws imposed on nature by the loving God who created it. No need for Jesus to save us: we're all on a Unitarian stairway to heaven.

Franklin helped turn Priestley's thoughts toward electricity, leading him to produce the first synthesis of the field and important new findings on the conductivity of resisters and the strength of electrical force. "What would Newton himself had said, to see the present race of electricians imitating in miniature all the known effects of that tremendous power . . . and amusing themselves at their leisure by performing with it all the experiments that are exhibited by electrical machines?" Priestley mused in his 1767 book *The History and Present State of Electricity.* To him, this was a brave new world of scientific understanding that informed his millennial theology. Later work in chemistry would cement his scientific reputation, but by the French Revolution, nothing could save him from being driven out of England for his leftist political and religious views, which his countrymen equated with radical antimonarchism and anticlericalism. Thomas Jefferson welcomed him to the United States, a republic where an Enlightenment era balance of powers and rule of law would supposedly protect his religious freedom.

Franklin, Wesley, and Priestley offer three different examples of how science related to religion during the Enlightenment in England and America: Deist, Christian, and Unitarian. Science may have been on the incline, and religion on the decline at that time in those places, but neither could be lightly dismissed. Both loomed large in the minds of people attuned to the latest intellectual currents, and many devised their own personal reconciliation of the two types of thought.

NINETEENTH-CENTURY DEVELOPMENTS

The utilitarian promise of electrical power materialized in the nineteenth century with the discovery and exploitation of current electricity. Detected at first through its stimulation of nerves in animals during the late 1700s, current electricity was initially seen as a material basis for life, leading some to raise religious concerns about the challenge it posed to spiritual accounts of human origins, and leading others, like Mary Shelley in her 1818 book *Frankenstein*, to denounce the hubris of scientists who would use it to play God. After Italian researcher Alessandro Volta established that an electrical current naturally passes between dissimilar metals separated by a moist conductor, and created the first voltaic pile, or battery, to show it, the race was on to discover uses for this constant energy source. For those with understanding of the science of and imagination about its applications, electric motors, dynamos, telegraphs, heaters, stoves, and arc lighting (if not a bulb) lay just around the corner. This was transportable power with limitless uses that would transform the way people lived and worked.

During the first half of the nineteenth century, no one understood electricity better or foresaw its applications clearer than the English natural scientist Michael Faraday. Humble to a fault and without formal education, but with a remarkable native insight into how electricity works, Faraday attracted the attention of Humphry Davy, England's leading chemical and electrical experimenter, and became Davy's assistant and successor at the Royal Institution of Great Britain, a premiere post for public science. In his mind, Faraday could virtually see the tensions and stresses in spaces caused by electricity and magnetism, and he drew on these insights to create a unified field theory relating and explaining those two previously distinct forces. Using his understanding of the underlying science, Faraday discovered or developed electromagnetic induction, electrolysis, diamagnetism, the electrical motor, and the dynamo. In 1845, he demonstrated that light was an electromagnetic phenomenon by rotating the polarity of light, passing it through a magnetic field. Nothing seemed beyond his reach in physics, and he received (and usually refused) every honor that an appreciative empire could bestow, including knighthood, which he declined because it violated his religious scruples against worldly titles.

Nothing in Faraday's science and none of his honors seemed to have had the slightest impact on his simple Christian faith. If physicists can compartmentalize their science and religion, then Faraday showed how. He was born into the church of his father—"a small and despised sect of Christians known, if at all, as Sandemanians," was how Faraday depicted it in a letter to Ada Lovelace in 1844—and never left it. Sandemanians worshiped as a loose alliance of working-class congregations without fixed creeds, trained ministers, or denominational organizations. They accepted a flat reading of scripture as their only foundational text and aspired to fellowship in the manner depicted in the New Testament. Faraday did so every Sunday, spending most of the day at church, often as an elder or deacon in his congregation, but devoted the rest of his week to work, and always insisted that he kept his science and religion strictly separate.

This was partly true. In conformity with his Sandemanian convictions, Faraday never tried to inform his religion by his science and eschewed all forms of natural theology. Human reasoning, he maintained, cannot lead to religious truth. Further, Faraday neither looked to scripture for scientific answers nor saw the Bible as a work of science. Nevertheless, at a fundamental level, his religious faith shaped his approach to doing science and technology. Faraday believed that nature was intelligible and useful because an infinitely wise God created it for people. "The Creator governs his material works by definite laws resulting from the forces impressed on matter," Faraday affirmed in an unpublished memorandum from 1844. "How wonderful is to me the simplicity of nature when we rightly interpret her laws." His goal became discovering those laws and their uses. He excelled at it. L. Pearce Williams wrote in his biography of this man, "That Michael Faraday, poor, uneducated son of a journeyman blacksmith and a country maid was permitted to glimpse the beauty of the eternal laws of nature was a never-ending source of wonder to him. . . . It was this attitude that earned Faraday his reputation for saintliness." It is a rare quality for physicists.

Faraday's genius was conceptual and descriptive, not mathematical, and ultimately modern physics relies on math. After all, Newton's *Principia* provided a mathematical description of gravity, not a mechanical one, and became the template for doing physics. Living a generation after Faraday and drawing heavily on his work, the Scottish physicist James Clerk Maxwell reduced Faraday's field models to mathematical equations in 1865, thereby establishing the classical theory of electromagnetic radiation that unifies electricity, magnetism, and light as three physical manifestations of the same natural phenomenon. A BBC survey in 1999 of one hundred top physicists listed Maxwell as the third-greatest physicist of all time, behind only Albert Einstein and Isaac Newton. Expressing his views on the subject, Einstein kept the portraits of three physicists on his office wall: Newton, Faraday, and Maxwell.

Despite differing from Faraday in background, training, and expertise—Maxwell was born rich, received the best scientific education available, and was a phenomenal mathematician—they shared a similar knack for compartmentalizing their science and their religion. And both remained true to the faith of their fathers till death, which in Maxwell's case meant to Scottish Presbyterianism and sustaining the country church near the family estate in Scotland after his father died. Still, like Faraday, he shunned public debates over science and religion, which grew ever more intense in Britain after the publication of Darwin's theory of evolution in 1859. Maxwell did not believe that science could prove religion any more than that scripture should be interpreted in light of modern science. In this he followed Kant, whose philosophy inspired the master of Maxwell's college at Cambridge, the influential English science polymath William Whewell. But much as for Faraday, a deep conviction that a created order underlay nature undergirded Maxwell's search for basic laws of physics. And unlike Kant, Maxwell privately wrote at university of subjecting his Christian beliefs to rational scrutiny and finding that they passed.

A public window into Maxwell's private views on the relationship between science and religion came from his popular writings and lectures on molecules. The first of these dated from shortly after Darwin's publication of *On the Origin of Species*. Others followed at irregular intervals nearly until the physicist's death at age forty-eight in 1879. Maxwell's remarks on molecules drew on spectroscopic analysis of light from distant galaxies and stars to argue that each type of atom and molecule in the universe is identical and unchanging across time and space yet displays a "manufactured" character that "precludes the idea of its being eternal and self-existent."

Even if life, the earth, and the cosmos have evolved, Maxwell was quietly affirming, in the midst of the deafening debates over the religious implications of Darwinism, that their basic building blocks—and the physical laws that underlie them—have not changed and reflect a designed order dating from the beginning of the universe. "No theory of evolution can be formed to account for the similarity of molecules, for evolution necessarily implies continuous change, and the molecule is incapable of growth or decay, of generation or destruction," Maxwell professed in an 1873 version of this argument delivered to the British Association for the Advancement of Science and published in the journal *Nature*. "We are therefore unable to ascribe either the existence of the molecules or the identity of their properties to any of the causes which we call natural." Thus, he declared, "We have been led, along a strictly scientific path, very near to the point at which Science must stop." And, he might add, where religion picks up: in the supernatural beginning of time and space.

A third prominent nineteenth-century British physicist, William Thomson, Lord Kelvin, was not so circumspect in using his science to defend his religion. Working first on the relationship between electricity and heat, Kelvin made his name during the 1850s developing James Prescott Joule's experimental findings on molecular motion as the source for heat into workable laws of thermodynamics. These laws provided a mechanical model for relating heat, mechanical work, and energy while giving direction to cosmic history. Although the energy in the universe is constant, Kelvin concluded, on average over time it dissipates irrevocably. Along with Newtonian mechanics and Maxwell's equations, thermodynamics provided a third leg for classical physics.

A devout Scots-Irish Presbyterian, Kelvin delighted in the thought that thermodynamics gave support to a theistic view of creation. Contrary to the deistic and agnostic uses of uniformitarianism in geology, which placed no limits on the earth's past or future, physics now showed a clear vestige of a beginning (in a total concentration of energy) and a certain prospect of an end (in the heat death of the universe). "A great reform in geological speculation seems now to have become necessary," Kelvin immodestly declared in his 1869 lecture "Geological Time." "British popular geology at the present time is in direct opposition to the principles of Natural Philosophy." This was religious apologetics pure and simple—using scientific arguments to defend religious doctrines—and even Kelvin labeled it as such. A brash man with a ready tongue, he also got it spectacularly wrong with air travel and X-rays, both of which he dismissed as impossible. He backed the losing direct-current side in the so-called war of currents until proponents of alternating current offered him a consulting contract. Over time, he became increasingly active in industrial research, apologetics, and conservative politics. Knighted for helping to lay the first transatlantic telegraph cable, Kelvin was made a lord for opposing Irish home rule.

After publication of *On the Origin of Species*, Kelvin turned his science against Darwinism by estimating the age of the earth through internal-heat loss and the age of the solar system based on how long a sun-size sphere of molten iron, silicon, or other known matter could emit enough heat to nurture life. Neither estimate was sufficient for the current array of living forms to have evolved through random mutations and unguided selection. "The limitations of geological periods, imposed by physical science, cannot, of course, disprove the hypothesis of transmutation of species," Kelvin said in his 1869 lecture, "but it does seem sufficient to disprove the doctrine that transmutation has taken place through 'descent with modification by natural selection.'" At the time, Kelvin's calculations offered potent scientific evidence against Darwinism. Even Darwin took note, and to accelerate the evolutionary process, he added to later editions of the *Origin of Species* doses of the Lamarckian methods of acquired characteristics through use, only to have Kelvin repeatedly shorten his estimates—ultimately settling on but twenty million years for the earth's age. Another solution, suggested by American geologist Thomas Chamberlin, among others, was that the sun and the earth had unknown energy sources—which were later found in atomic fission and fusion.

With Kelvin, who lived into the twentieth century, we have come full circle from Enlightenment-age French encyclopedists who used physics to promote atheism. "If you think strongly enough you will be forced by science to the belief in God," Kelvin affirmed in a widely reprinted 1903 response to a popular lecture by Christian apologist George Henslow. "You will find science not antagonistic but helpful to religion." Yet Kelvin did not speak for physics in 1900 any more than d'Alembert did in 1750. Although they offered individual examples of the relationship between science and religion as seen by two well-known physicists, both men represented extreme cases. And as much as each thought their era represented something of a culmination of progress for their profession, the twentieth century overturned classical physics and opened new ways of relating science and religion.

RELATIVITY, QUANTUM MECHANICS, AND BEYOND

We introduced the topics of relativity and quantum mechanics in our chapter on astronomy, but there is somewhat more to say about them now that we are talking about physics. During the early twentieth century, relativity and quantum theory launched the universe of modern physics that still rules the roost today, though one might more fairly say that they enriched (rather than replaced) Newtonian mechanics, electromagnetic field theory, and thermodynamics. Certainly they threw theological speculation about the religious meaning of physics for a loop and, as noted in the last chapter, opened new ways of thinking about how God might work in the universe. In an effort not to repeat ourselves too much, three main points should suffice here: indeterminacy, complementarity, and holism.

But before addressing these points, let's review what we've covered. Classical physics envisioned an utterly deterministic and totally reductionist universe that was objectively knowable through science. Newton had left room for God in this system by observing non-reductionist instability in planetary orbits that would require ongoing divine intervention and by pointing to the solar system's unnatural orderliness as evidence of a designer. By 1800, however, Pierre-Simon Laplace had accounted for these matters and proclaimed the theoretical ability to wind the entire system forward or backward with perfect precision, in a manner that not only dispensed with the need for but disproved the possibility of ongoing supernatural intervention in nature. God may or may not exist—and people might still claim the ability to feel God's presence spiritually, and that feeling might be true—but God was banished by classical physics from the ongoing operation of nature, although not necessarily from its origin in time and space. Human free will fared no better under determinism.

Einstein's view of God fit the classical formulation. "I

cannot conceive of a God who rewards and punishes his creatures," Einstein wrote in his essay "The World as I See It," yet he claimed "a firm belief, a belief bound up with deep feelings, in a superior mind that reveals itself in the world of experience." This, he said, "represents my conception of God." Belief in an impersonal, rational, objective God led Einstein to believe in a real, knowable, deterministic universe. It drove his search for a grand unified theory of everything and spawned his objections to the indeterminacy of quantum mechanics. Popular interpretations of Einstein's theories of relativity sometimes posit that they deny objectivity in nature, whereas they actually address the objectivity of the observer.

Later generations of theoretical physicists working on relativity, whose numbers include Steven Weinberg and Stephen Hawking, continue the quest for a grand unified theory in much the same manner as Einstein did. To discover it, Hawking wrote in his best seller from 1988, *A Brief History of Time*, "would be the ultimate triumph of human reason—for then we would truly know the mind of God." Hawking later denied that this was a religious statement—he was still an atheist, he assured his readers—yet it captured the essence of Einstein's God and the quasi-religious passion driving modern physics.

In a sense, there is nothing much new here for the science-and-religion dialogue. Dutch philosopher Baruch Spinoza could have said much the same about God in the seventeenth century, and probably could have said it better than either Einstein or Hawking. When combined with the quantum mechanics that repelled Einstein, however, relativity opened new pathways. Indeterminacy attracted the most attention.

Quantum mechanics grew out of the discovery that, unlike how it was perceived under classical physics, energy is released and absorbed in tiny packets called "quanta." While this appeared to make little practical difference at the normal sensory level for humans, it could make a huge difference at the molecular level, which in turn could have real-world impact (at least in theory, as backed up by mathematical models).

During the 1920s, German physicist Werner Heisenberg and Austrian physicist Erwin Schrödinger derived some surprising implications from all this. At the quantum level, and therefore to some extent at any level, Heisenberg found that the position and momentum of quantum-level particles cannot be known simultaneously—and since they can't, then the universe is not predictably deterministic. The observer must choose which to measure, and by choosing, fixes one and makes the other less knowable. Further, since quantum-level phenomena such as light exhibit a wave-particle duality, Schrödinger showed that multiple measurements of them taken at any one time would find the probability of their location within a wave rather than at a single point. By cracking the deterministic causality of classical physics and putting the observer into the observation, indeterminacy opened the possibility of undetectable divine intervention in nature and human free will, leading the Quaker astrophysicist and popular science writer Arthur Stanley Eddington to comment in his Gifford Lectures in 1927, "Religion first became possible for a reasonable scientific man about the year 1927."

This may have been an exaggeration. There is no evidence of an uptick in religiosity among scientists since the 1920s, but Eddington's comment drew cheers in theological circles, even though most of those shouting the loudest probably had no idea what Heisenberg's uncertainty principle meant. But if uncertainty at the quantum level left room for God to act and humans to choose, then those already believing in a proactive God and human free will would take it to heart and pass along the good news. Certainly, believing parents would tell it to their children as they headed off to study physics at Cambridge, Chicago, or Berkeley. One of us, our historian who taught in Georgia for twenty years, knew a brilliant engineering professor at Georgia Tech who had the argument down pat, but he was an evangelical Christian whose faith seemed to rest more on biblical certainty than on Heisenbergian uncertainty.

Then there was the metaphor of complementarity. Niels Bohr, the Danish founder of quantum mechanics, introduced the principle of complementarity to correlate the information about a quantum-level object obtained by separate, mutually exclusive experiments. One experiment might show light to be a wave, while another, done at the same time, would have shown it to be a particle. Such findings, Bohr said, were complementary: two ways of knowing. Neither was more correct than the other, and both added information about the object under investigation. Further, he added, complementarity in knowability implies complementarity in an object's essential nature. Reading Bohr, some theologians adopted this metaphor for scientific and religious ways of knowing. A religious account of transcendental meditation, they might say, could complement (rather than contradict) a scientific account of it, in that both might be true in some sense and add to a fuller understanding of it. Returning the compliment, when ennobled by Denmark's king, Bohr put the Taoist symbol for yin and yang on his coat of arms, along with the motto in Latin, "Opposites are complementary." It did not mean that he had found religion, however.

Holism in quantum physics refers to how individual subatomic particles within a single physical system relate at a fundamental level. When two such particles with large momentum collide at a known position, for example, measuring the resulting momentum of one will simultaneously resolve the momentum of the other and render the position of both less knowable. Yet this involves the instantaneous passage of information between two receding bodies, which violates the rules of classical physics and relativity. Bohr attributed such interactions, which subsequently were detected experimentally, to the holistic nature of the physical system that incorporated both the objects and the instruments measuring them. Holism had many implications for religion. It challenged the reductionism of classical physics, which saw the behavior of the whole as reducible to the behavior of its material parts, by showing that those parts only behave as they do as part of the whole. Further, it offered a physical parallel for how God might have an active presence everywhere at once. Finally, by stressing the impact that observers have on observations, it suggested that individual choices can make a difference. All that happens need not be predetermined. The object, subject, and even God may grow out of the process.

THE TAO OF PHYSICS

Between us, we have taught in universities for over sixty-five years and have never seen anything like the philosophical response to quantum mechanics for any development from science, though perhaps the buzz surrounding genetics in the 1990s and neurobiology today comes closest. During the interwar years from 1918 to 1939, theoretical physics became the darling of academics, philosophers, theologians, and intellectuals alike, with a beguiling allure that reached into popular culture. On virtually a daily basis, its leaders, particularly Einstein and Bohr, seemed to be redefining the nature of reality. Their words carried weight and were repeated widely. Both men became international celebrities, though Einstein, with his unruly hairdo and witty remarks, gained notoriety most of all. Bohr's Institute of Theoretical Physics in Copenhagen became the fountainhead of a movement, with tentacles reaching out in Europe to Berlin, Rome, Paris, and Cambridge, and from there to American research centers in Cambridge, Chicago, Princeton, Berkeley, and Caltech, which became the places for astrophysics. A second tier of academic institutions spread the work around the globe. The best science students flocked to these top centers like aspiring actors to Hollywood.

Interaction with religion became especially intense. During the 1920s, quantum mechanics directly impacted the development by Alfred North Whitehead of process philosophy, which focuses on the interconnectedness of all things, including God. But there was more. Progressive, open-minded Eastern mystics, Islamic scholars, Jesuit priests, Christian ministers, and Jewish rabbis claimed parallels between modern physics and their traditions. Buddhists quoted Heisenberg's observation in *The Physicist's Conception of Nature:* "Even in science, the object of research is no longer nature itself, but man's investigation of nature," as harmonious with their spiritual awareness. And the feeling was mutual. "For a parallel to the lesson of atomic theory," Bohr wrote in his book *Atomic Physics and Human Knowledge*, researchers should look "to those kinds of epistemological problems which already thinkers like the Buddha and Lao Tzu have confronted, when trying to harmonize our position as spectators and actors in the great drama of existence." These quotes date from later, but similar sentiments were expressed at the time. Although he called himself an atheist, Schrödinger displayed strong interest in Eastern religions. By the 1930s, in the wake of one world war and with another looming ahead, as people looked for understanding during the Great Depression, the new physics appeared to offer as good a source for it as any, and better than most.

When that second global conflagration came, physics went to war. Einstein and Schrödinger fled their homes in Germany and Austria. Relativity was damned by Nazi scientists—including two with Nobel Prizes in physics—as Jewish. After Germany occupied Denmark, Bohr's institute went quiet. Heisenberg turned to war projects for the German Army Weapons Bureau. Émigré physicists did the same for the American and British military, side by side with the best and the brightest home-grown scientists. Einstein urged the United States to build an atomic bomb, which became the largest wartime project of Allied science. In the end, the deadliest conflict of all time became known as "the physicists' war" and fittingly closed with the destruction of Hiroshima and Nagasaki by "the physicists' bomb." The bloom was off the rose.

Awed by its power, postwar physics became the spoiled

ward of governments, with the United States and the Soviet Union lavishing funds on favored researchers. Suspicions of disloyalty could bring retribution, however, as it did on the brilliant theoretical physicist who had headed the American atomic bomb project, Robert Oppenheimer. And what better way to arouse suspicion in Cold War America than to read *Siddhartha* or refer to Eastern religious thought in the manner of prewar quantum physicists. Idealistic young scientists increasingly turned in other directions, particularly toward biology, where they saw more chance of helping people. Edward Teller, the émigré physicist who supposedly inspired the title character in the 1964 movie *Dr. Strangelove*, became the face of Cold War physics. Whatever his attributes as a scientist, and they were legion, Teller was unlikely to kindle religious interest in physics except perhaps as a diabolical force. An agnostic, in his *Memoirs* Teller said of his Jewish upbringing in war-ravaged Hungary, "The idea of God that I absorbed was that it would be wonderful if He existed: We needed Him desperately but had not seen Him in many thousands of years." It took hippies to save physics for religion.

American psychologist and LSD advocate Timothy Leary allegedly saw it coming. "There must be thousands of young persons whose nervous systems were expanded and opened-up in the 1960's and who have now reached positions of competence in science," he is quoted by MIT historian David Kaiser—in his book *How the Hippies Saved Physics*—as saying in 1977. "We expect the new wave of turned-on young mathematicians, physicists, and astronomers are more able to use their energized nervous systems as tools to provide new correlations between psychology and science." The Fundamental Fysiks Group, based in the Bay Area and loosely linked to the University of California's Berkeley Center for Theoretical Physics, was part of this wave. Openly taking psychedelic drugs, practicing transcendental meditation, and drawing inspiration from Eastern religious concepts of holism, members of this group took up Bohr's concept of quantum entanglement and its offshoot, Bell's Theorem, to generate the No-Cloning Theorem, which became a key to quantum computing and the basis for quantum encryption in communication.

The Fysiks Group's greatest impact for science and religion came through its role in reviving a perceived linkage between modern physics and Eastern mysticism. Group member Fritjof Capra's book from 1975, *The Tao of Physics*, became a runaway best seller and New Age classic. Other books followed, a virtual subgenre of literature exploring the meaning of organic interconnectedness and holism from scientific and religious perspectives. "Quantum theory and relativity theory—the two basis theories of modern physics—force us to adopt a much more subtle, holistic and 'organic' view of nature" than classical physics, Capra wrote in his book, and to experience the universe as "a dynamic, inseparable whole which always includes the observer in an essential way." Observer becomes participator, he stressed, just as in Hinduism, Buddhism, Chinese thought, Taoism, and Zen—each of which he presented in separate chapters but whose essential elements he blurred into a single form of mystic spirituality. "These elements also seem to be the fundamental features of the world view emerging from modern physics," Capra affirmed.

As with any effective inspirational book, Capra concluded *The Tao of Physics* with a call for repentance and transformation. "I see science and mysticism as two complementary manifestations of the human mind," he wrote in the book's epilogue. "Neither is comprehended in the other, nor can either of them be reduced to the other, but both of them are necessary." Calling for "a radically different social and economic structure" compatible with environmental sustainability and human betterment, Capra closed by saying that, "to experience the wholeness of nature and the art of living with it in harmony," humans will need to balance the yang of science with the yin of mysticism. "The survival of our whole civilization may depend on whether we can bring about such a change." In Capra's view, science and religion were complementary ways of knowing and equally essential for humanity. Modern physics told him so. In this view, however, he remained an outlier among modern physicists. Most would more likely agree with the ever quotable Nobel Prize–winning American physicist Steven Weinberg, when he wrote in his popular book *The First Three Minutes*, in 1977, "The more the universe seems comprehensible, the more it also seems pointless." An atheist, Weinberg repeated his take on the relationship between science and religion in 2006 at a Sauk Institute for Biological Studies conference in San Diego with the words, "The world needs to wake up from its long nightmare of religious belief; and anything that we scientists can do to weaken the hold of religion should be done." So much for either complementarity or complexity. For Weinberg, conflict reigns.

CHAPTER THREE

The Brain, the Mind, and the Soul

HAVING concluded our discussion of modern physics and religion with the holistic view that the observer inevitably impacts the observation, and, earlier on, noting the Cartesian impulse in physics to explain all aspects of life in mechanistic terms, this is as good a place as any to introduce issues raised by these lines of scientific thought for religious conceptions of the human mind and soul. If the brain (as part of the body) is merely machinelike matter in motion, as René Descartes supposed in the early seventeenth century, and if the mind or soul are simply manifestations of the brain, as he most decidedly did not, then the physics, chemistry, and other scientific aspects of the human body and brain necessarily impede on religious concepts of the mind and soul. And as these issues are philosophical as well as historical, we will let our philosopher take the lead—beginning as he did in the first chapter with the ancient Greeks.

THE MIND OF PLATO

The common-sense position on the mind-body relationship—what one might call the fallback position or the null hypothesis—is some form of dualism, meaning that body and mind are two different things. Every human has a physical body and a mental world—a kind of external world and a kind of internal world, although no one thinks that if researchers dissected a fresh human brain or, worse yet, a living one they would find some sort of identifiable mind in there, much less a soul. That is the whole point. The mind and, by implication, the soul are different from the body, and thus are different from the brain.

One of the earliest versions of this sort of position is to be found in the dialogue the *Phaedo*, written by the Greek philosopher Plato. His thinking on the subject is set against an overall metaphysical picture that supposes that the physical world is in some sense merely a reflection or shadow of absolute reality, the world of Ideas or Forms. These Forms are eternal, unchanging archetypes, not physical but in some sense "rational," meaning that their existence is something to be apprehended by the intellect, very much in the way that people apprehend the objects of mathematics, like right-angled triangles. The *Phaedo* tells of the last day on earth of Plato's teacher, Socrates, who had run afoul of the authorities for filling young men's heads full of subversive ideas, and was about to be forced to drink poison. Those with him are amazed at his calm demeanor and wonder why he is not afraid. He explains that what they see and what will die is his physical body. His mind, his mental body, that which in some sense exists in the world of Forms, since it is able to apprehend them, is like the Forms eternal and will never die. Thus there is no cause for fear.

This dualistic position (and a somewhat modified version of it held by Plato's pupil Aristotle) has been incredibly influential through the history of Christianity, primarily because it was picked up in late antiquity and endorsed by the most influential Christian thinker of them all, Augustine. He interpreted the Genesis passage that speaks of humans being made in the image of God as meaning that humans have minds, which in some sense he identified with souls, which make them thinking beings—rational and able to discern good and evil. Science, he suggests, is eternal: "For what exists and is unchangeable [obviously Augustine has mathematics in mind here] must be eternal." And this means that since the mind is able in some sense to internalize the knowledge of science, "the human mind always lives."

In some manner or other, this kind of thinking is held by many (one suspects most) Christians today. To take but one modern-day example, the British philosopher Richard Swinburne, contributing to a symposium "Thinking about Death" (2004), was forthright in his subscription to such a view of the body and mind. "The concept of me is the concept of a soul," he wrote, "and since undoubtedly I exist and so there is something satisfying that concept, I am essentially a soul, an immaterial substance distinct from my body. Hence, at death, when the soul is separated from the body, it is possible that it continues to exist. So if there is a God, he can, if he so wishes, make it continue to exist either on its own or by connecting it again to its old body or to a new body, if its previous body has been annihilated." Note incidentally the range of possibilities in this last sentence. Plato and, following him, Augustine rather suggested that immortality is going to be a soul-only sort of existence. The biblical author and first-century Christian apostle Paul, it should be noted, operating more in the Jewish tradition than a Greek one, did not think of the soul existing without the body. Indeed, he did not think much about souls. Humans are going to be resurrected in some wise after death, Paul believed, but for him this meant the resurrection of the body in some glorified form, presumably with brains attached.

THE CARTESIAN MIND

Skipping over the medieval theological debate about the relationship between the mind and the body, the issue is placed center stage in science during the Scientific Revolution by the seventeenth-century French philosopher and mathematician René Descartes. By then, as noted in chapter 1, a mechanical metaphor had replaced an organic one to explain physical things, and Descartes was totally committed to a mechanical view of the world, thinking of it (and its contents) as something akin to those marvelous church clocks that not only told the time but had mechanisms to show the movements of the heavens, and figures striking bells (to mark time), and so forth. Presenting the human body as if it were mere robotics, in his *Discourse on the Method* (1635) he wrote:

> This will hardly seem strange to those who know how many motions can be produced in automata or machines which can be made by human industry, although these automata employ very few wheels and other parts in comparison with the large number of bones, muscles, nerves, arteries, veins, and all the other component parts of each animal. Such persons will therefore think of this body as a machine created by the hand of God, and in consequence incomparably better designed and with more admirable movements than any machine that can be invented by man.

The heart, which to Aristotle had been the seat of human and animal emotion, became to Descartes merely a teakettle pump that warmed the blood and sent it out through the arteries and back through the veins mechanically. He had more trouble devising a mechanical analogy for the brain but never doubted that one existed. Only the mind or soul stood apart. For animals—even one's beloved pet dog, Descartes noted—it was purely mindless mechanics.

Cartesians combined this kind of thinking with an argument made famous by the slightly later German philosopher and mathematician Gottfried von Leibniz, who (in his *Monadology*, 1714) invited his readers to join with him in entering a mill—basically a piece of machinery so large that people could get inside it. "It must be confessed that perception and that which depends upon it are inexplicable on mechanical grounds, that is to say, by means of figures and motions. And supposing there were a machine, so constructed as to think, feel, and have perception, it might be conceived as increased in size, while keeping the same proportions, so that one might go into it as into a mill. That being so, we should, on examining its interior, find only parts which work one upon another, and never anything by which to explain a perception." Today this is known as the "hard problem of consciousness." Back then it was known as "Leibniz's gap" and it is easy to see why. After investigating all sorts of machinery and studying its workings, there is still nothing more than material things in motion. Thinking or consciousness is just not there, or if it is there, it does not stand out as something distinct from the material brain.

For a Christian like Descartes, therefore, since humans clearly have the physical machine body and because equally clearly they are beings with minds—Descartes is most famous for proving human existence with his claim "I think, therefore I am" (*Cogito ergo sum*)—body and mind must be two distinct substances. These Descartes referred to as *res extensa*, "extended substances," and *res cogitans*, "thinking substances." In this sense, Descartes viewed humans as part physical or animal and part spiritual or divine. Like all other animals, humans have *res extensa*, which accounts for their physical bodies and which has only temporal existence; like God, humans have *res cogitans*, which accounts for their spiritual soul and which presumably has eternal existence. For Descartes, this was as much a scientific theory as a religious belief—with both parts having their scientific and theological critics. For example, the Catholic Church censored the idea. Thus, the underlying puzzle of material brain and thinking mind remained for science and religion. Two particular historical approaches to it merit further discussion due to their ongoing relevance.

THE MIND OF DARWINISM

First, there is Charles Darwin's mid-nineteenth-century theory of evolution by natural selection, which remains the dominant paradigm in the modern life sciences. We will speak much more of Darwin and his theory later, but strictly with respect to the mind, Darwin argues that, for humans, the mind is right there at the heart of the evolutionary process. In his private, often cryptic, notebooks back in 1838—twenty-one years before he published the landmark *On the Origin of Species*—the very first intimation he gives that he has grasped the idea of natural selection as a force for change is a passage talking not just of human beings, but of their mental abilities: "An habitual action must some way affect the brain in a manner which can be transmitted.—this is analogous to a blacksmith having children with strong arms," he mused in a Lamarckian fashion, but then added, "The other principle of those children, which chance? produced with strong arms, outliving the weaker ones, may be applicable to the formation of instincts, independently of habits."

Humans, including their brains, are products of evolution by natural selection, Darwin is suggesting. But does this include such human mental attributes as consciousness? The philosopher Daniel Dennett asserted that consciousness has been fully explained by science without any reference to God or divine souls and, although he put cognitive science up front, he included Darwinian evolutionary theory in the explanatory mix. But it cannot be denied that there are some, perhaps many, leading researchers who would counter that Darwinism fails to explain consciousness. In *How the Mind*

Works (1997), the noted Harvard-based evolutionist Steven Pinker wrote, "At least for now, we have no scientific purchase on the extra scientific ingredient that gives rise to sentience. As far as scientific explanation goes, it might as well not exist." But as Pinker went on to point out, there is much that scientists can do with sentience once they have it on the table. So let us, or at least the philosopher between us, go on to suggest what can be said about Darwinism and mind and about the implications from this for religion.

Darwinism—evolutionism generally—stresses the continuity and interrelatedness of all life. Humans are not entities alone. They are part of the seamless tree or web of life. For a start, humans are animals in their environment as much as cats and pigs and snakes and fish and ants and worms. In the notebook passage immediately after the one just quoted, Darwin grasped this and the stupendous implication that, if humans have minds because they have brains, then one can hardly deny some kind of mind for those others with brains. "Can we deny relation of mind & brain. Do we deny the mind of a greyhound & spaniel . . . ?" he asked. This cuts right into what the French seem to find a natural and acceptable implication of Cartesian dualism, and what the British, for nigh four centuries, have found revolting and insensitive about such dualism. For Descartes, body machines can function autonomously. Humans have minds also, but this is as much for theological reasons as anything. Logically, they could have been zombies with just machine bodies, unthinking, going through the motions. And this is indeed the case for all animals other than humans. They have no connection with *res cogitans*. They are simply unthinking machines. Hence, practices like vivisection on animals, which seem to involve unutterable pain on the subjects, are quite acceptable because no thinking beings are suffering. The British have generally agreed with the philosopher David Hume that animals have passions no less than humans. They may not be as fully conscious as humans, but they have minds of some degree and kind.

The implications of all of this for religion are obvious and immediate. The Abrahamic religions certainly elevate humans above other animals. People uniquely are made in the image of God. Yet there has always been a concern at some level for the rest of life. Humans do not own it. They are its custodians, set in dominion over it. Precisely what that means is a matter of some debate, but it does at least mean that there is a bond, and Darwinism supports this. (More on this point in chapter 9.) For Eastern religions like Buddhism, with reincarnation absolutely central, Darwinism is perhaps even more important. If animals are just unthinking machines at the lower levels, as a punishment there is little point in being reborn as one. Nobody is going to take that situation as an opportunity to change and start moving upward. Without some kind of awareness in one's lower animal state, the theology collapses. Cartesian dualism—Descartes himself was a sincere Catholic—is the philosophical equivalent of prussic acid for any religion that has a significant place for all living beings.

So much for the positive case. Is there any more to be said, perhaps showing that the result of Darwinian science looking at the mind is to point people away from the tenability of theism in particular and religious belief in general? One who certainly thought so was Darwin's most prominent late-nineteenth-century supporter Thomas Henry Huxley. He argued that although Darwinism certainly does not prove the non-existence of mind, it does show it to be froth on the top. A full-blown epiphenomenalist, he claimed that humans are fundamentally machines with cloudlike minds sitting uselessly on top. In the striking language for which he is still famous, in an essay "On the Hypothesis that Animals Are Automata, and Its History" (1874), Huxley wrote, "The consciousness of brutes would appear to be related to the mechanism of their body simply as a collateral product of its working, and to be as completely without any power of modifying that working as the steam-whistle which accompanies the work of a locomotive engine is without influence upon its machinery. Their volition, if they have any, is an emotion indicative of physical changes, not a cause of such changes."

Obviously if Huxley is right, then a fairly hard-line form of materialism follows. Just as obviously, if it is true, the world of science becomes hostile to religion—a conclusion that Huxley welcomed. When the body goes, the mind goes, and even if the body is recreated, it is far from obvious that it would be the same body or the same mind. More significantly, perhaps, the very thing that theists count as really significant—the view that conscious rational beings are made in the image of God—has been downgraded to ephemera. It is no more than a useless by-product of the physical brain.

Whether people have to accept this kind of thinking is another matter. The American pragmatist and no-less-ardent Darwinian William James thought not. He agreed that it is possible that consciousness and the associated reasoning could be useless by-products on the top of the functioning brain-mechanism. The fact is, however, that complex biological phenomena, especially those that seem to be doing something, tend not to be by-products. It is true that in their origins they may just have started to appear; but if they can serve any function, natural selection very quickly picks up on them and modifies them or intensifies them or some such, turning the new item to the advantage of its possessor. And it is hard to deny that reasoning and consciousness do have adaptive value. They help people survive. As such, consciousness has real, meaningful existence, but under Darwinian theory, that still does not necessarily make it anything more than an epiphenomenon of the physical brain.

COMPUTERS AND THE BRAIN

Perhaps because they are such an omnipresent aspect of life in the twenty-first century, computers may now impact how people see the mind-body dichotomy more than Darwinism did. For example, around 1980, Herbert Simon, who won a Nobel Prize in economics including his work in artificial intelligence, used to include in his talks a little reflection about computers. Imagine, he would say, passengers twenty years ago (1960) are on board a plane and the pilot announces that it is being flown by a computer. They would, Simon said, reach for their parachutes and jump. Simon would then bring the audience up to date (1980), saying: imagine the pilot announces that the computers have all died. Once again passengers would be reaching for parachutes.

Simon drew attention to the huge revolution that occurred in the second half of the twentieth century, when machines for thinking—or, if one insists (for this goes to the heart of our discussion), "thinking"—were devised and rapidly became so powerful and dependable that they moved right into all facets of modern life, basically transforming the technological underpinning on which people now depend utterly and completely. And that was in 1980! Reflect on how computers have developed since, most particularly with laptops and other accoutrements like sophisticated phones. Can anyone under the age of thirty imagine a world without Facebook? Can anyone over thirty remember a world without Amazon.com? Above all, can virtually anyone survive without texting? These things have become part of modern existence.

It is not surprising that, part cause and part effect, a whole new discipline has grown up on and around computers—so-called cognitive science. It is new and yet it is not that new, for it is a logical extension of issues raised in the past two chapters. As physicists and other researchers have explored the world of experience through the metaphor of the machine, science has gone from strength to strength since the great changes of the sixteenth and seventeenth centuries. Now, just as people have used their brains to build computers, so computers are being used to understand and explore human brains. Indeed, it is affirmed that the brain is a computer, although as one of the leaders in the field, Marvin Minsky from MIT, has noted, it is a funny kind of computer—a computer made of meat.

From this, people have been quick to draw conclusions that impinge directly on the relationship between science and religion. The philosopher Andy Clark, in *Mindware: An Introduction to the Philosophy of Cognitive Science* (2001), wrote:

> This notion of the brain as a meat *machine* is interesting, for it immediately invites us to focus not so much on the material (the meat) as on the machine: the way the material is organized and the kind of operation it supports. The same machine can, after all, often be made of iron, or steel, or tungsten, or whatever. What we confront is thus both a rejection of the idea of mind as immaterial spirit-stuff and an affirmation that mind is best studied from a kind of engineering perspective that reveals the nature of the machine that all that wet, white, gray, and sticky stuff happens to build.

This is strong stuff, and it is not just one person's idiosyncratic high-flown talk. Turn to Daniel Dennett's runaway best seller *Consciousness Explained.* After giving a full and careful account of the brain as a physical computer, he confidently asserted that any sense of the spiritual is undermined. Humans need no more information now that they know the mechanism of consciousness. Mind, soul—call it what one wills—is explained—or, more precisely, explained away. This is a bit too extreme for most people, since computers cannot yet simulate consciousness, but there are many who share Dennett's general perspective while thinking that perhaps a little more work (both scientific and philosophical) is needed to get to the same conclusion that matter in motion is all there is; there is no spirit world outside material brains of humans. Descartes was right all along, this view concludes, except for the soul part. Like their pet dogs, humans—body, brain, and mind—are simply *res extensa.* And as for God, well, everyone knows that "god" is just "dog" spelled backward.

COMPUTERS: FRIEND OR FOE OF RELIGION?

The biblical book of Genesis affirms that humans are made in the image of God, which is usually taken to mean that people have minds that can reason and make moral decisions plus, for many, that they have a spiritual, supernatural side as well as physical, material existence. It is a central element of Christian theology that physical death is not the end, but that in some sense people have the prospect of ongoing existence. That is where the talk of immortal souls comes in and some sort of spiritual existence, though the Bible itself does not mention souls as something distinctive when God was doing His creating.

Generally people do not want to identify the mind exactly with the soul, whatever that might mean, but equally generally people do want to connect the two. Humans have minds and souls, most theists believe, stones and plants have neither. And obviously if the two can be connected, and if the mind in turn is nothing but a manifestation of a physical machine—a computer made of meat—then at once there are significant theological or religious implications. It would seem that when the body goes, the brain goes, and when the brain goes, the mind goes, and when the mind goes, the soul goes. Short of some kind of physical resurrection of the body and the brain, there are going to be no minds or souls enjoying (or hating) eternity.

It seems from this that, as Dennett presents it in *Consciousness Explained*, cognitive science is no friend of Western religions that view humans as having eternal souls, and, in respects, it is pretty obvious that it is an even worse friend of Eastern religions. Take Buddhism, a religion that has reincarnation—the transmigration of souls—at its center. Essential to the theological ontology of Buddhism is that there are layers of existence. Humans are in a middle level. There are those below them and those above them. At the bottom level is the hell realm, *niraya*, with vile beings tortured and subject to horrible nightmares. Then above this comes the level of *petas*, ghostlike creatures, bearing analogy to the phantom spirits of Western lore. Next up is the animal realm, sharing physical space with humans but in major respects these are (and have the defining features of) lower forms of life. Humans sit above these but below one or two levels for the gods—the *asuras*, the lesser gods, and the *devas*, which include the *brahmas*, the very highest form of being.

In addition to this, Buddhism suggests that there are an infinite number of universes, with galaxies, themselves clustered into thousand-fold groups. There are innumerable planets with inhabitants, much like Earth and its denizens. Everything is subject to *dukkha* (suffering)—although, because there is no Creator God, and especially no Creator God who was good and had in mind the special status of humans, there is no one to blame for suffering. It just happens. Humans can choose to do good or ill, and depending on this they direct their future assignments and their levels. For the Buddhist moving toward enlightenment, this is done with the aim of getting off the process and achieving *nibbana* (nirvana), although whether this is simply nothingness or something more—an existence that is endless and wholly radiant, the "further shore," the "island amid the flood," the "cool cave of shelter," the "highest bliss"—is a matter of some debate among believers.

Yet the kind of radical materialism that cognitive science supports undermines such beliefs. If the mind and the soul are not just connected to the brain but simply a function of brains, then it is hard to see how one links one existence to another, since clearly the material brain does not transport. None of it makes any sense at all under Dennett's *Consciousness Explained* or any similar understanding of cognitive science.

It seems therefore that in this area of modern science people are plunged straight into the science-religion conflict, and religion is going to be left on the battlefield, beaten to a bloody pulp and dead. But is this really so? Do religious believers have to buy so quickly into the conclusions drawn by the likes of Daniel Dennett and other cognitive scientists—that if the brain is fully explained as a machine, then there can be no spiritual element to humans? Historically, two notable solutions have been offered.

One is to grasp the nettle and maintain that once separate, always separate, no matter what computers show. Body and mind are two and never will be one. The two substances work in parallel, never interacting, not even in the brain, but never getting out of synch. This position, known as "psychophysical parallelism" is associated with the seventeenth-century French priest and philosopher Malebranche, as well as with Leibniz. It is argued that God does the heavy lifting, keeping the system going smoothly. Which is all very well but doesn't really speak to its implausibility. Why would God set up two separate systems and then not connect them but superintend them both, all of the time?

But is the other solution any better? Here one drops the dualism and argues that there is only one substance and that both body and mind are manifestations of that one and the same. "Monism" escapes the problem of dualism because there was no problem in the first place. There is nothing to be separate from something else. Note that this is not quite the position of Daniel Dennett. For him in some way mind is an "epiphenomenon" of the brain, meaning that the brain is the basic substance and mind somehow comes out of this, inasmuch as there is something, for Dennett, to come out in the first place. For the monist—the seventeenth-century Dutch philosopher Baruch Spinoza was one such thinker; the nineteenth-century German evolutionist Ernst Haeckel was another—there is one substance, and both body and mind are equal manifestations of this. The problem now, of course, is how anyone makes sense of all of this. Is it plausible? The common move is to make a distinction between sense, or meaning, and reference. The morning star and the evening star seem to be two different things, and people can go through life never putting them together. But they are in fact manifestations of the planet Venus. The reference is Venus, the meanings are morning and evening star. Perhaps this helps the body-mind case. The reference is the underlying substance; the meanings are brain and mind. The trouble now is the same as before, however, and somehow the gap between body and mind seems just as great.

NEW MYSTERIANISM

At this point, many people throw up their hands and cry "cave." They sense that this is a problem without a solution, or at least without a solution open to human intelligence. This is less a move of desperation or cowardice, but more a conviction that there simply isn't a solution of a kind that people could appreciate; they don't even know what a solution would look like. In respects, because of advances in science, especially advances in the physical sciences teaching that some questions simply cannot be asked because there are no possible answers available, this is perhaps a more attractive option in recent years than it was back at the time of Descartes. Quantum mechanics particularly bars physicists from inquiring too deeply into the wave-particle nature of the electron or the exact moment in genetics when one molecule mutates into another. Perhaps the body-mind problem is like this.

It goes almost without saying that this position—known today as "new mysterianism"—has its attractions for theists. Christianity notoriously stresses that humans cannot know everything—"Now we see through a glass, darkly" was how Paul put in the Bible—and that they should not expect to grasp fully what it means to be made in the image of God. "Apophatic" theology—trying to define God and God's effects negatively from what God is not—has always had its critics; but it (and the associated mysticism that senses the divine but cannot fully articulate the experience) has a long tradition in Christianity. The early theologian Tertullian in his *Apology* (about 200 CE) wrote, "That which is infinite is known only to itself. This it is which gives some notion of God, while yet beyond all our conceptions—our very incapacity of fully grasping Him affords us the idea of what He really is. He is presented to our minds in His transcendent greatness, as at once known and unknown." Today, elements of such thinking can be found far and wide, including in the writings of that runaway favorite of the modern-day evangelical Christians, C. S. Lewis. In *Letters to Malcolm* (1964) he wrote: "For our abstract thinking is itself a tissue of analogies: a continual modelling of spiritual reality in legal or chemical or mechanical terms." Continuing: "Are these likely to be more adequate than the sensuous, organic, and personal images of scripture—light and darkness, river and well, seed and harvest, master and servant, hen and chickens, father and child? The footprints of the Divine are more visible in that rich soil than across rocks or slag-heaps." It is little wonder that Lewis turned to writing fictional stories about a fantasyland, Narnia, because in a way he thought that these could convey deeper insights than anything in literal science or philosophy or theology.

Interestingly, this position might also find favor with Buddhists. It has always been a key part of Buddhist thought that there are questions that human reason cannot answer and that indeed people should not really ask. The Buddha objected that spending too much time on metaphysical issues detracts from the tasks at hand, in particular facing and overcoming suffering, and the pursuit of *nibbana.* The fourteen "unanswerable questions" are divided into four categories: about the existence of the world in time; about the existence of the world in space; about life after death, specifically about the existence of the Buddha after death; and, of obvious interest here, about the nature of personal identity. Is the self, the mind, identical to the body, or is it something else? Far from new mysterianism being imported into Buddhism, it is right there at the core of the religion. That it took Western thinkers so long to come to this kind of thinking is less a matter of wonder and triumph and more one of regret and pity.

BACK TO THE PRESENT

So having completed this survey of the mind-body issue in the history and philosophy of science, what has been learned? A great deal. It is remarkable to discover the extent to which brains are—or at least can be simulated by—computers. Already they can beat the best of the best humans at chess. Already they are more accurate at diagnoses than the most skilled of physicians. Already they are showing their prowess at successful applications to college. And one can only assume that they will get better and better. At one level there is nothing mysterious about organisms, and that includes human beings. The mechanists were right. Descartes is vindicated. Humans are machines, which means that any adequate religious perspective must start right there. Claims about this world must be at the very least compatible with this fact. For instance, if it is claimed that morality is important, then it must be possible for organic machines to be programmed to make their possessors moral.

So, again, having completed this survey of the mind-body issue in the history and philosophy of science, what has been learned? Nothing much. The problems of Descartes and his successors about the implications of humans-as-machines remain. For all of the confident claims of Daniel Dennett and others, Leibniz's gap still looms large. A machine working—mill or computer—does not as such imply consciousness. Some have suggested that invoking a hardware-software distinction avoids this problem. The brain is the hardware; the mind is the software. But the analogy doesn't really work. A computer program and the stored files aren't really conscious.

There are a couple of fairly recent, much discussed thought experiments that make the kind of point being made here. The pioneering computer scientist Alan Turing, back in 1950, proposed a test for thinking. If a computer could answer questions accurately enough that no one could spot anything amiss, then there would be no reason to deny that it is a "thinking" machine. But of course the critic is going to object that simply going through the mechanical motions is not thinking. To think otherwise is to beg the question à la Dennett. In 1980, the philosopher John Searle proposed as a counter his famous "Chinese room" objection. Suppose someone with absolutely no knowledge of Chinese were in a closed room with suitable texts and manuals, and that questions in Chinese were posted through a slot in the wall. Using the manuals, that person might answer the questions and post out correct answers, having no knowledge at all of what the questions or answers meant. For example, suppose she received the question (written in Chinese) "What color are lemons?" She looks down the pages of her texts until she finds the question and then looks across to where it says there are answers and then she writes down the Chinese answer, which the manual tells her is the Chinese word for "yellow." Even if she is doing things accurately enough to pass the "Turing test," no one would say that there was genuine thought involved. A machine could have done what she did, and indeed today would probably do it much faster and more accurately.

So in a way, here we are back where we started. Some people might, despite all of the problems, opt for dualism by believing on faith that humans have machine brains but also non-machine minds, souls, or spirits. Computers apart, this approach might be slightly more plausible today than it was back at the time of Descartes. One thing that modern physics has taught is that the idea that the world is made up of inert chunks of basic substance, matter, is simply not true. Whatever may be the case down at the quantum level, there certainly are not simply minichunks of rock there. Matter is energized—dare one even say "alive"—in a way not dreamed of even in the nineteenth century. This certainly doesn't prove dualism; but it does mean that mind and matter may not be quite as far apart as they might have seemed to Descartes. However before one grabs it happily and takes up with Augustinian notions of the soul, do note that if people like William James are correct, matter and mind are a nigh-inseparable pair. They work intimately together. At the least this makes some kind of parallelism implausible; there doesn't seem to be a third element, even God, between matter and mind.

Alternatively, as discussed earlier, religious believers can take refuge in the new-mysterianism view that not only has no one yet solved the body-mind problem, but that it is beyond solution. Dennett thinks that cognitive science refutes this position already, and others are optimistic that if not today, then tomorrow. Many disagree. The challenge here, as much as anything, is to know even what a solution would look like. Take an analogy: The problem of the origin of life has not yet been solved and may never be solved adequately. But researchers know what a solution would look like—energy from deep-sea rifts, an all RNA world, and so forth. So, there is reason for science to keep pursuing the problem to the end. In the case of the body-mind problem, however, there is reason to say that, in some fundamental respects, scientists are no further along than philosophers were at the time of the *Phaedo.* Obviously scientists know a lot more about consciousness and how it is connected to the brain than anyone knew in Plato's day. But the hard parts of the problem remain.

CHAPTER FOUR

Rock, Fossil, God

ONCE upon a time people took the earth for granted. Depending on where they lived, they might have felt it quake or erupt, but these were exceptions. To most people at most times, the earth seemed like a constant that needed little explanation. Sure it served as the stage for the great drama of life, but who notices a stage? For non-creationist religions, such as Hinduism or Buddhism, the earth is just here. For creationist ones, such as Christianity, Judaism, and Islam, God had created the earth, but, at least after Adam's fall and the flood of Noah's time, it has stayed the same. Since people didn't see much change in the land around them—and they are not good at seeing processes longer than their life spans—they pretty much took the earth at face value until closer scientific examination began raising questions about the scriptural accounts. Indeed, most early modern European geological writers, such as Danish naturalist-turned-bishop Nicolas Steno and Italian abbot-naturalist Anton Moro, relied on Genesis as an infallible source while making lasting contributions to the understanding of rock formation. By the early 1800s and thereafter, however, geologic discoveries forced many serious-minded Christians either to reappraise their understanding of scripture or reject key tenets of mainstream scientific theory. With our historian taking over from our philosopher, this chapter chronicles these developments both for what they reveal about the interaction of science with religion and to accelerate the transition from focusing on the physical sciences to considering the life sciences.

ACCOUNTS OF CREATION

The biblical account of creation appears in the book of Genesis, which is sacred scripture for Jews and Christians. The Qur'an incorporates or references this account in three places, making it foundational for Muslims as well. For conservatives among these three great Western faiths, this account represents the revealed word of God and, as such, carries special meaning in some literal, topologic, allegoric, or mystic sense. Even for liberals with these traditions, the Genesis account carries meaning as an early record of the Jewish people's understanding of God's role in creation. Accorded any of these meanings, the Genesis account was (and is) a document to be reckoned with in understanding origins.

The first chapter of Genesis tells of God creating the heavens and the earth, then plants and animals, and finally humans—all in six days. Significantly, all types of plants and animals are said to reproduce "according to their kind." Read literally, this precludes evolution from one "kind" of plant or animal to another and suggests that, since God presumably wouldn't make anything for naught, all these different kinds are still around. Regarding humans, the account declares that God separately created them in God's own image and likeness. The second chapter of Genesis contains an alternative account in which the order of the appearance of life forms on earth is somewhat reversed—but with a similar emphasis on the special creation of humans by God. Indeed, it is this second account that first introduces Adam and Eve as the progenitors of the human race, with God directly forming them as man and woman.

Similarly, the Islamic Qur'an speaks of God creating Adam as the earth's "viceroy." Adam then names the created animals and is placed by God with Eve as his wife in the Garden of Eden where they are tempted by Satan, a fallen angel, and sin. Most extensively in Surah 11, but in at least eleven other places as well, the Qur'an speaks of the prophet Noah and the great flood sent to destroy the people who would not listen to his call to repentance. Unlike in Genesis, however, the Qur'an does not depict this flood as worldwide, and it is harder to see it as destroying all life outside the ark or fundamentally transforming the landscape.

Neither the Bible nor the Qur'an explicitly state when these creation events happened. Most early Christians and Jews probably assumed that they all happened within the generations of named individuals listed in the Bible, going back to Adam. During the mid-1600s, Anglican archbishop James Ussher used these generations to calculate the year of creation as 4004 BCE, or less than three thousand years before Genesis was supposedly written by the Hebrew leader Moses. Printed in the margins of the Authorized, or King James, Version of the Bible, Ussher's chronology became quasi-gospel for British and American Protestants during the eighteenth and nineteenth centuries. Jews, Muslims, and other Christians were not bound to Ussher's chronology but likely held a similar basic view of the earth's past. For all of them, the earth was created relatively recently and, with two possible exceptions, had not changed much.

In the Hebrew scriptures as in the Qur'an, the first exception occurs shortly after creation when Adam and Eve sin and God casts them out of the idyllic Garden of Eden. Known as the Fall in Christian theology, this episode allows for dramatic changes in the world. At least for Christians and Jews, if not for Muslims, the second exception appears later in Genesis when God causes a great flood or "deluge" to sweep over the entire earth, killing all animals (or at least all land animals) other that those taken by Noah and his family on the ark. Since this reportedly included breeding pairs of all kinds, the originally created types presumably survived. Under this view of creation, fossils could not tell us anything about past species, because none of them need have gone extinct. Geologic features, either as originally created or as reconfigured by the flood, also seemed constant.

Generally speaking, Christian leaders from the early church through the Reformation did not view the Bible as a scientific text. Science, in the sense of a distinct intellectual tradition seeking naturalistic explanations for physical phenomena, began with ancient Greek philosophy roughly five hundred years before Christ. Although most individual Greeks probably accepted religious or mythical explanations for natural phenomena, some Greek philosophers sought to banish the supernatural from the natural by proposing purely materialistic accounts of nature. Nothing is aught but physical matter in meaningless motion, the ancient Greek atomists proclaimed. The origin of life and individual species posed a particular problem for Greeks intent on devising purely materialistic explanations for natural phenomena. Creation implies a Creator, and so to dispense with the need for a biological creator, such ancient philosophers as Anaximander, Empedocles, the atomists, and the Epicureans advanced various crude notions of organic evolution.

Based on his study of animal anatomy, however, Aristotle concluded that species are absolutely immutable. Each species always breeds true to its form, he maintained, and never gives birth to an evolutionary type. Rejecting both creation and evolution, Aristotle (a non-theist) simply posited that species are eternal. Integrating the Genesis account with Aristotelian thought, premodern Christian theologians and naturalists typically viewed each species as created by God in the beginning and thereafter fixed for all time in a perfect (albeit fallen) creation. Well into the nineteenth century, most Western naturalists saw little scientific reason to reject Aristotelian thinking on the fixity of species—and fully appreciated the religious advantages of retaining it. Whether read literally or not, the Genesis account comports with the idea that species don't change, fossils reveal little about the past, geology is fixed since the flood, and nothing on earth is very old.

BURSTING THE LIMITS OF TIME

Cracks began appearing in this comfortable worldview with the coming of the Enlightenment of the 1700s. Copernicus and company had already burst the limits of space by greatly expanding the perceived size of the cosmos and removing the earth from its center. The revolution in astronomy fed a wider enlightenment of elite understanding that sought rational explanations for politics, economics, society, the rest of nature, and even religion. Among eighteenth-century French scientists, Georges-Louis Leclerc, Comte de Buffon, personified the Enlightenment. One of the foremost descriptive naturalists of his day, Buffon was also a highly original theorist who, as superintendent from 1739 to 1788 of the King's Garden in Paris, commanded the position and prestige to promote his novel ideas about nature. Although historians still debate whether Buffon was strictly a deist or some less-radical form of theist, he certainly rejected Christianity and sought naturalistic explanations for the origin of the earth and its inhabitants.

Those explanations punctuated his massive treatise, *Natural History*, which appeared in fifteen initial and seven supplemental volumes over the period from 1749 to 1789. Earth and other planets congealed from globs of molten matter thrown off when a comet crashed into the sun, Buffon proposed in volume 1, and living things spontaneously generated on the earth as it cooled, he added in later volumes. For proof, he offered crude experiments with molten iron balls, whose cooled surfaces conveniently wrinkled like the earth's terrain, and with boiled meat gravy, which became alive with microorganisms when cooled. Although Buffon never offered a precise timetable for these events, presumably they extended much farther back than a Genesis chronology permitted. Capturing the spirit of the era, Buffon's speculations inspired a generation of researchers. By disrupting traditional beliefs, they may have also contributed in some small way to the revolutionary fervor that swept Paris shortly after Buffon's death and replaced France's Catholic monarchy with first an atheistic republic and later a military dictatorship under Napoleon.

No French naturalist of the next generation had a greater impact on modern biological and geological thought than Georges Cuvier. Born a Lutheran in a remote duchy subsequently annexed by France, Cuvier was inspired to study science by reading Buffon, and in the early 1790s he attracted the attention of a scholar who had fled the terror of revolutionary Paris. "I have just found a pearl in the dunghill of Normandy," the scholar wrote about Cuvier to a colleague who had remained in Paris. After the terror subsided and old royal establishments reopened as republican institutions, the young provincial found himself in Paris and quickly rising in the ranks of professional scholars at Buffon's old King's Garden, which expanded into Paris's grand National Museum of Natural History.

Concentrating his research on the comparative anatomy of higher vertebrates, Cuvier became convinced that the internal structure of animals revealed their function and therefore their true nature. For him, form followed function. His research profited greatly from his position at the world's premiere natural history museum—an institution that rapidly became ever more comprehensive in its zoological holdings as Napoleon's armies plundered the collections of Europe and sent home live, preserved, and fossilized specimens from as far afield as Russia and Egypt. Ultimately, Cuvier proposed that there are four (but only four) basic anatomical types of animals: vertebrates (with backbones), mollusks (with shells), articulates (like insects), and radiates (like starfish). This view, built solidly on anatomical analyses, shattered the hierarchical concept dating from Aristotle of a single chain of beings rising in fine gradations from the simplest living form to humans at the top.

Cuvier was the first naturalist to have at his disposal a suitably complete collection of the world's mammals—past and present—to make definitive distinctions among them. He made the most of this advantage. In 1796, for example, Cuvier announced that, based on his anatomical comparisons of actual specimens, the elephants of India and Africa constituted two distinct species and that both of them differed from the larger elephantlike mammoth found only in fossil remains. The positive identification of other living and extinct mammals followed, one after another in rapid succession. To account for so many extinct species, as early as 1796 in his essay "Memoir on the Species of Elephants," Cuvier announced "the existence of worlds previous to ours, destroyed by some kind of catastrophes." And with ever closer study of the fossil record, that earlier world soon became a succession of geologic epochs stretching into the indefinite and seemingly infinite past. Although he remained Lutheran at a time and place where many French intellectuals were deist, Cuvier was forging the solid scientific evidence for shattering the Genesis chronology of a single recent creation of the earth and it inhabitants.

Before Cuvier, European naturalists typically held that no species—perfect in original creation—ever die out. Fossils had no fundamental significance to them: such things were simply sports of nature or remnants of some still-living species. Overturning this view, Cuvier ultimately concluded that *all* fossilized animals differed in kind from modern ones and that *no* modern species existed in truly fossil form. In the essay titled "Preliminary Discourse," Cuvier boldly claimed the power "to burst the limits of time, and, by some observations [of fossils], to recover the history of the world, and the succession of events that preceded the birth of the human species."

Suddenly, life had a history different from the present, and fossil fragments revealed it. "As a new species of antiquarian," Cuvier explained in "Preliminary Discourse," "I have had . . . to reconstruct the ancient beings to which these fragments belonged; to reproduce them in their proportions and characters; and finally to compare them to those that live today." The science of paleontology was born in Cuvier's laboratory. Because of his conviction that the form of any animal precisely served its functional needs, Cuvier confidently assumed that trained researchers could reconstruct its entire structure from any one of its functional parts. Paleontologists could do for extinct animals what comparative anatomists did for living ones—definitively identify them. Doing so for all of the earth's past and present species became Cuvier's goal for science—and he would launch the effort, doing his own best work with fishes and four-footed mammals.

Living in a particularly volatile era of French religious history characterized by alternating phases of Enlightenment deism, revolutionary atheism, and Restoration Catholicism, Cuvier stood apart from most others within the cultural elite of France by remaining a churchgoing Protestant during his entire life. Indeed, he visibly aligned himself with his religious minority by overseeing government programs for Protestant education and serving as vice president of the Protestant Bible Society of Paris. He married a socially prominent Roman Catholic widow of the Terror, Anne-Marie Coquet de Trayzaile, but they raised their children as Protestants. When his daughter, Clementine, adopted an evangelical form of Protestantism, however, she grew to doubt her father's salvation and prayed for his conversion.

It was not about to happen, at least on her terms. By definition, evangelicals publicly proclaim their religious beliefs and seek to convert others to them. For Georges Cuvier, however, religion was a strictly private matter of personal faith. Perhaps it had to be so for him to prosper in French science and politics, but that gives an unjustifiably cynical slant to the matter in Cuvier's case. Although he was the very embodiment of reason in his science and in public, Cuvier accepted religious truth as existing wholly apart from reason. This made his private religious beliefs virtually invisible to others, even to his own daughter, and they have remained so to this day despite considerable speculation. Yet surely he was a Bible-believing Christian of some sort, and these beliefs inevitably impacted his science. Indeed, they helped to make his radical reconception of geological history acceptable to Protestants scientists and theologians in Europe and America.

BIBLICAL DAYS AND GEOLOGIC AGES

For Cuvier, the geologic column suggested a historical pattern of catastrophic floods—perhaps six in all—alternating with periods of terrestrial uplift. These floods were massive deluges, he surmised, such as the world has not seen at least since the biblical time of Noah. Each of these floods caused mass extinctions, scoured the terrain, and laid down a layer of sedimentary rock containing the fossil remains of the former population. When the land resurfaced in a later geologic epoch, different kinds of plants and animals repopulated it—presumably through migration from other regions, but perhaps of a new creation. Cuvier's writings suggest both options without clearly excluding the seemingly supernatural second one. At the most, he wrote in his *Historical Report on the Progress of Geology Since 1789*, "One is authorized to believe that there has been a certain succession in the forms of living beings" as shown by fossils in the geologic column.

Although Cuvier did not know when the species originated, he was certain that countless types had become extinct, which was news enough in 1800. This discovery came from establishing beyond a reasonable doubt that that fossilized animals found in the older layers of the geologic column differed in kind from living ones. "It is the generality of this difference that makes it the most remarkable and astonishing result that I have obtained from my research," Cuvier proclaimed in the essay “Extract from a Work on the Species of Quadrupeds of which the Bones have been Found in the Interior of the Earth.” “I can now almost assert that none of the truly fossil quadrupeds that I have been able to compare precisely has been found to be similar to any of those alive today.” As his investigations progressed, Cuvier further generalized this assertion to cover various types of animals from different geologic epochs. For example, he later commented in “Preliminary Discourse” on marine mollusks: “The shells in the ancient beds have forms that are specific to them, and . . . are no longer found in the recent beds. Still less are they found in present seas, where the analogues of their species are never discovered, where even many of their genera are not found.”

The geologic column did show a progression of forms from mostly simply ones toward the bottom to more complex ones toward the top, Cuvier observed. The types of animals also became increasingly like present-day ones over time. For example, he continued his comment on mollusks by noting, “The shells of the recent beds, by contrast, resemble in their genera those that are alive in the seas; and in the last and least consolidated of these beds, there are some species that the best-trained eye cannot distinguish from those that the ocean [now] sustains.” To some, such evidence suggested evolution. Cuvier had already rejected this explanation based on his study of comparative anatomy by concluding that each type is too irreducibly complex to change, and the apparent absence of transitional forms in the fossil record confirmed this conclusion. In his extensive study of fossils, Cuvier saw only distinct species that persisted without change until they went extinct altogether at some remote time in the earth's unimaginably long history.

In his basic conception of natural history, Cuvier carried mainstream scientific opinion with him. In fact, his rhetorical appeal to facts and repudiation of theorizing proved most persuasive to conventional, conservative naturalists—precisely those scientists most likely to accept traditional viewpoints, including biblical ones. After a generation of battering by radical French materialists, some Christian intellectuals (particularly in Britain and the United States) welcomed the findings of a prominent French naturalist whose views were not openly hostile to their own. They met him at least halfway.

In 1813, a year after its French publication, the pious Scottish geologist Robert Jameson translated Cuvier's popular "Preliminary Discourse" into English under the title *Essay on the Theory of the Earth*, with a preface and notes stressing points where the French naturalist's views coincided with Christian doctrine. Working in the shadow of French rationalism, Cuvier never invoked scriptural authority to support his scientific arguments, but his "Preliminary Discourse" did observe that the date traditionally ascribed to the biblical deluge that supposedly drowned all life outside Noah's ark roughly coincided with geologic evidence for the time of the last catastrophe. Jameson's annotations trumpeted this observation, the repeated seemingly extranatural catastrophes, and Cuvier's devastating critique of evolution. It should be noted that the French naturalists who followed Cuvier did not necessarily see these past catastrophes as extranatural, and Jean-Baptiste-Jacques Élie de Beaumont would attribute to them the quite natural shrinking of the earth, as over the vast expanse of geologic time, these nonreligious explanations were lost on some Bible-believing Christians in search of scientific support for their faith.

During the first half of the nineteenth century, other conservative Christian geologists in Britain and America labored to reconcile the new geologic orthodoxy with the Genesis account. In 1814, Scottish natural theologian Thomas Chalmers proposed that a gap existed in that narrative between the book's first and second verses. This opened unlimited time for geologic epochs between "the beginning" and God's creation of current species. Amherst College geologist Edward Hitchcock adopted this so-called gap theory and popularized it in the United States. Meanwhile, Scottish geologist Hugh Miller suggested that the days of creation in Genesis symbolized geologic epochs or ages. Yale geologists Benjamin Silliman and James Dwight Dana (a team of father-in-law and son-in-law) championed the "day-age theory" in America, with Silliman backing geological ages and Dana favoring cosmic ages.

Cuvier's followers modified his basic outline of geologic history to keep abreast of the latest scientific findings during the mid-nineteenth century. For example, to avoid choosing between a single creation of all life and multiple creations following the various catastrophes, Cuvier maintained that migration could account for the abrupt appearance of new species in the local fossil record of a particular place. As wider fieldwork gradually eliminated plausible sources for the migrating species, however, many of Cuvier's followers turned to multiple mass extinctions and creations as the most realistic explanation for the abrupt appearance of species in the fossil record under a creationist model.

Cuvier's equation of the biblical deluge with the final catastrophe lost its principal scientific proponents in the 1830s, when British geologists Adam Sedgwick and William Buckland, both evangelical Christian Oxbridge dons, concluded that a single flood of the type described in Genesis could not produce the complex deposits attributed to the last catastrophe and should have left human fossils, which were never found among its debris. Yet they drained the biblical deluge of geologic significance without drying up their Christian faith: the ages of creation simply moved back in time, more in line with the biblical account that places the days of creation before Noah's time. Also by the 1830s, Cuvier's Swiss-American disciple Louis Agassiz had shown that ice ages (rather than floods) likely caused the mass extinctions found in the fossil record. With such revisions, Cuvier's old-earth catastrophism thrived well into the mid-nineteenth century. For some, it became a model for how modern scientific empiricism could coexist with traditional religious belief by interpreting the Bible to fit the facts of science while steering clear of the wild theorizing characteristic of pre-Darwinian evolutionists such as Lamarck.

NEPTUNISM AND VOLCANISM

Even as Cuvier was giving life to the remote past through paleontology, German mineralogist Abraham Werner was devising the first comprehensive theory of ancient geologic history. Like Cuvier, Werner did so without stirring religious controversy, even as he pushed the earth's age far beyond any traditional notions associated with the Genesis account. Where Cuvier's insights came from studying fossils, Werner was seeking to understand how and where minerals were laid down so that his students could exploit them for mining, manufacturing, and agriculture. He was so successful that many Europeans and even some Americans went to study under him at the Freiberg School of Mines from 1775 to 1817, and carried his Neptunist theory of an ancient earth back home with them. At the very least, the increased mining activity that accompanied the Industrial Revolution gave a boost to geology by making its study immensely practical.

As Werner saw it, most rocks and geologic formations were created through the gradual retreat of a mineral-rich primordial ocean that once covered the entire globe. Granite precipitated out of this vast sea first, followed by basalt, minerals, crystals, and the various sedimentary rocks. Werner dismissed volcanoes as the products of burning coal seams that produced localized lava flows. To account for fossils and other complexities, he had the water rising and falling even after the first life forms appeared, but he never even tried to explain where it all came from or went. This grand aquatic process took time, Werner stressed, perhaps a million years or more. He did not equate this massive global ocean to the biblical deluge, but others readily did—utterly ignoring problems of chronology. Further, Werner's geologic history, like the biblical account of the earth's past and future, had a beginning, a direction, and a probable end. A handful of pious American students took this reading of geologic history back home, where they introduced it into their own teaching at colleges in the United States, most of which had church ties. For his part, although raised in German Pietism, Werner adopted a romantic, quasi-transcendentalist Unitarianism without objecting to other readings of his work.

Resistance to Werner's geology came less from Christian theologians than from volcanist-minded naturalists who stressed a continuing and undiminished role for the earth's inner heat in rock formation, such as the late-eighteenth-century Scottish polymath James Hutton. On philosophical grounds, Hutton could not accept Werner's world. A radical deist, he believed that God surely created a self-sustaining earth that suffered no permanent directional change yet was continually refreshed for human life by ongoing, currently observable forces. Like Werner, Hutton may also have been influenced by the Industrial Revolution swirling around him, particularly his Scottish compatriot James Watts's remarkable steam engine, which went around and around driven by internal heat.

In his concept of steady-state volcanism, which satisfied both of these constraints, Hutton proposed a cyclical process of igneous-rock mountains gradually rising from the earth's molten core and then slowly weathering to create inhabitable land. As this land accumulated over time, its bottom layers would push down toward the fiery core and remelt. The pressure resulting from accumulating layers would push up new mountains from the molten core, and the deliberate, law-bound cycle would repeat. "The result, therefore, of our present enquiry," Hutton, in his 1788 paper "Theory of the Earth," famously declared, "is, that we find no vestige of a beginning, no prospect of an end." Despite the burden of Hutton's obtuse prose and idiosyncratic religious views, and the accumulating evidence of direction in the fossil record, steady-state volcanism survived as a minority view that tempered the prevailing catastrophist, directional tone of early-nineteenth century geologic thought. Then in 1830, the English lawyer and gentleman geologist Charles Lyell gave it wings.

Called to the bar in 1822, Lyell grew bored of practicing law and opted to make his name in geology—a subject that had fascinated him since he took Buckland's courses as a student at Oxford. Lyell consciously and with apparent sincerity chose to champion a geologic theory that was at once revolutionary enough to excite interest yet compatible with his elevated social status and Unitarian religious beliefs. He found that theory in Hutton's steady-state volcanism, which Lyell updated with fossil and other evidence to create the so-called modern theory of uniformitarian geology.

Before Lyell, mainstream scientific opinion gave a "directional" orientation to geologic history. The earth was hotter and wetter in the distant past, both Werner and Cuvier maintained, and past geologic events were more dramatic than present ones. Living for the most part in the now geologically quiet regions of northern and western Europe, they could not conceive of current geologic forces shaping the earth's features and causing discontinuities in the fossil record. Those forces must have been greater in the past. By Lyell's time, geologists in the directionalist camp had a naturalistic explanation for the diminishment of geological forces over time that also relied on the earth's inner heat. Recent empirical evidence had found that the earth did indeed possess an inner heat, and the laws of physics suggested that was diminishing over time. Drawing on this, some directionalists argued that the earth had cooled and contracted during its long geologic history, with the rate of that cooling and contraction gradually decreasing until the current era of relative stability when solar radiation offset interior heat loss. Historian of geology Martin J. S. Rudwick has explained in *Earth's Deep History* and elsewhere that the theory of a cooling earth implied that conditions on the surface and in the earth's crust would have been very different in the earliest epochs from what they later became. This could account both for geological formations and the fossil record suggestive of seemingly catastrophic past changes without invoking supernatural intervention. The earth and its inhabitants were different in earlier geologic epochs, and naturally so.

Schooled in the directionalist geology that he dismissed as "catastrophism," Lyell embraced steady-state volcanism with a convert's zeal and transformed it into modern uniformitarianism. In his three-volume *Principles of Geology*, first published in 1830–1833, Lyell refashioned Hutton's cyclic outline of geologic history into a coherent scientific theory. Using the polemic skills of a barrister to make his case, Lyell wove in observations from Italy and other volcanically active or earthquake-prone regions to suggest that (even at current rates but given limitless time) the earth's inner heat could dramatically sculpt geologic features. He also turned the fossil record to his advantage by stressing that the breaks in it were neither so complete nor so dramatic as directionalists claimed. Quite to the contrary, he argued (on the basis of a few disputed examples) that representatives from all classes of plants and animals appeared throughout the fossil record.

Indeed, although he agreed with directionalists that God specially created species to fit their environment, Lyell saw long-term environmental change as gradual rather than abrupt and therefore posited that new species were created continuously rather than in spurts. An epoch predominated by mammals could just as easily precede as follow one predominated by reptiles, Lyell believed; it simply depended on environmental conditions at the time. Of course, the fossil record did not fully display a cyclical pattern of life, but Lyell attributed this to its incompleteness. Fossils are only laid down in particular conditions, he noted, and eventually destroyed in the cyclical subsidence of older rock strata. While Lyell always stressed the unique place of humans in creation, seeing us as the direct product of the divine hand, he otherwise challenged the notion of progress over time and, until Darwin persuaded him otherwise in 1860, denied the possibility of organic evolution.

PROGRESS IN THE FOSSIL RECORD

Through his books and lectures, Lyell became a force within the growing and influential British scientific community. In part it was a matter of method, as scientists increasingly became less willing to hypothesize unobserved and untestable past events of supernatural magnitude to explain physical phenomena, and more willing to acknowledge the role of currently observable geologic forces in shaping the earth's features. For Darwin, uniformitarianism greatly lengthened the time available for evolution to operate and illustrated the cumulative power of small changes. But Lyell made little headway in eliminating the profession's or the public's perception of progress in geologic history. Widespread interest in fossil discoveries made that virtually impossible. This requires backtracking a bit to bring developments in paleontology up to Lyell's heyday.

Where Hutton and Lyell saw cyclic, directionless change in the fossil record, the directionalist camp saw progress. This became a fault line within natural history during the period. To most naturalists, Cuvier's findings suggested a progression of fossil forms, with fish and reptile fossils in earlier rock strata, mammal fossils added in later ones, and human remains found only amid the most recent geologic deposits. Overlapping with Cuvier's early work on these matters, the English geologic cartographer William Smith found while cutting cross-country canals for British industry that he could definitively identify each strata of sedimentary rock by the characteristic mix of the fossil species that it contained. Smith's findings reinforced and extended the concept of progressive organic succession. And as paleontologists made similar findings in the fossil record of more places, the conclusion became increasingly inescapable to them by the 1820s: new species both appeared and disappeared over time, and the mix seemed to grow more complex with time. To some anthropocentric naturalists, the order of appearance seemed distinctly progressive in that it led up to humans. In a rough sense, this concurred with creationist religious opinion.

With traditional beliefs about a single creation and the permanent endurance of created species shattered, interest in paleontology soared. For the first time, fossils had meaning. They represented unknown past life forms containing clues to geologic history and biologic origins. By finding new fossils, nineteenth-century Europeans and Americans could participate at the cutting edge of Western science. Interest died down as the fossil record became better known, but for a season it captured the popular imagination. Among those newfound fossil species, none attracted more attention than dinosaurs.

Of course, people had found dinosaur fossils before, but they meant little to someone who did not believe in extinction or who did not identify them as coming from dinosaurs. To such a person—scientist or not—they were merely big bones rather than remnants of lost giants. Once naturalists accepted the idea of extinct species, and especially after they gained some notion of an "age of reptiles" prior to the appearance of mammals, then the "discovery" of dinosaurs fired the scientific and popular imagination. People began seeing them for what they were, and it changed their view of geology and paleontology.

Cuvier identified the first great reptile of a bygone era from a four-foot fossil jawbone captured by the French Republican Army during its sweep of the Meuse region in the Netherlands in 1795. Known as *Mosasaurus* (lizard of Meuse), this was a huge aquatic reptile from secondary chalk beds near Maastricht. These much-quarried ancient beds contained a rich array of fossil invertebrates, fish, and reptiles, but no mammals. Cuvier later used them to illustrate his observation that reptiles appeared much earlier than mammals. He identified *Mosasaurus* as an extinct lizard with anatomic similarities to present-day monitor lizards of the tropics but much bigger and solely aquatic. The honor of identifying the first giant prehistoric land reptiles (though they too did not at first classify them as "dinosaurs") fell to an incongruous pair of English naturalists, William Buckland and Gideon Mantell.

Buckland was a boisterous, pre- and early-Victorian polymath. Ordained as an Anglican cleric, Buckland was tapped in 1818 for a new Oxford readership in geology, a subject that he'd loved since his childhood. He rose to the top of the British scientific establishment over the next three decades while holding church posts ranging from country priest to dean of Westminster Cathedral. Despite his prestigious positions, Buckland never took himself (or his colleagues) too seriously. His flamboyant lecturing style became legendary at Oxford. To illustrate his lectures on dinosaurs, for example, he might lumber around the lectern mimicking the gait of an overstuffed land lizard or flap his clerical coattails like a winged pterodactyl. Known for attributing the odd collection of prehistoric animal remains found in a Yorkshire cave to its supposed use in warmer prediluvian times as a hyena den, Buckland kept a live African hyena at his home. He also had a pet bear, who accompanied him to college functions wearing a cap and gown. Spoofing his reputation for discovering fossil reptiles, he served alligator meat to favored guests. As a young man, Charles Darwin found Buckland's antics off-putting and in his autobiography attributed them to "a craving for notoriety, which sometimes made him act like a buffoon"—but then Darwin went to Cambridge.

An avid collector of natural-history specimens, Buckland obtained his first dinosaur fossils from a slate quarry near Oxford during the 1810s. "The detached bones," he later wrote in the scientific paper "Notice on the Megalosaurus," "must belong to several individuals of various ages and sizes. . . . Whilst the vertebral column and extremities much resemble those of quadrupeds, the teeth show the creature to have been oviparous, and to have belonged to the order of Saurians or Lizards." Their size was striking. One thighbone measured nearly three feet long and ten inches around. If its relative proportions matched those of living lizards, then this land animal had "a length exceeding 40 feet and a bulk equal to that of an elephant seven feet high," Buckland reported in the Megalosaurus paper. It took some time for him to figure out what he had found, but it came from secondary strata roughly the same age as the one that held Cuvier's *Mosasaurus* and, emboldened by that precedent, in 1824 Buckland finally published a description of his *Megalosaurus* (great lizard), the largest land animal heretofore identified.

By the time that Buckland published his discovery, Gideon Mantell had found even larger fossil remains of the same species in southern England. "The beast in question would have equaled in height our largest elephants, and in length fallen but little short of the largest whales," Buckland speculated in his Megalosaurus paper. Mantell practiced medicine for a living but lived to collect fossils—making his best finds in a sandstone quarry near his home in Sussex. His *Megalosaurus* fossils came from this site in 1821.

In 1822, Mantell found an enormous fossil tooth, worn down like those of a plant-eating mammal. Yet mammal fossils did not come from this stratum. Even Cuvier was stymied by the tooth, at first identifying it as from a rhinoceros and only later writing to Mantell, "Might we not have here a new animal, a herbivorous reptile?" Mantell confirmed Cuvier's suggestion by comparing the tooth with those of present-day iguanas, and found a match in everything except size. His tooth would have come from a sixty-foot-long iguana, Mantell estimated. He published his find in 1825, making his *Iguanodon* (iguana tooth) the second great terrestrial reptile of record.

Mantell identified a third such species of giant prehistoric reptile, the armor-plated *Hylarosaurus* (woodland lizard), eight years later from a remarkably complete fossil embedded in a block of Sussex limestone. In 1841, these three fossil species, by then confirmed in multiple specimens, became the founding members of Dinosauria, a newly minted prehistoric suborder of saurian reptiles. The British public loved it and flocked to displays.

Buckland and Mantell labored to fit their dinosaurs into a broad temporal context, and for both of them, that context took an even more decidedly progressivist turn. They were Cuvierian directionalists, to be sure, which instinctively led them to see progress in the fossil record. Cuvier had confined his analysis to two basic divisions of sedimentary rock—the older Secondary strata, rich in fossilized fish and marine reptiles, and the younger Tertiary strata, where the fossils of land mammals first appear in great numbers. He never presented this as a progression of types over time. Although Cuvier ducked the issue, the discovery of large land reptiles in Secondary strata disrupted the simple equation of earlier times with aquatic animals: the presence of terrestrial dinosaurs implied the existence of vast pre-Tertiary continents. In an 1831 paper, for example, Mantell placed his dinosaurs in a "geological age of reptiles," in which giant lizards ruled the sea, land, and sky. It followed an age of fish and preceded an age of mammals, a seemingly progressive sequence.

Befitting his academic and ecclesiastical status, Buckland offered a grander and more directional scheme than Mantell and gave it a religious imprimatur. Directionalists had long seen the earth as formerly hotter and wetter. Some naturalists, including Buckland, saw this climatic change as a proximate physical cause for organic succession. As the climate changed, he reasoned, a good God would create different sorts of plants and animals to fit it. A thoroughgoing Christian directionalist, Buckland envisioned his God creating a progressive succession of species, each perfectly designed for the climate of its geologic epoch and all pointing toward the ultimate creation of humans in God's image when conditions became suitable.

Of course, Buckland's view of biologic history drew on biblical revelation as well as paleontological evidence, but not in a strictly literalistic manner. From the outset, Buckland posited a gap in the Genesis account that left ample time for a long geologic history before the creation of current forms, and he gradually recognized a diminished role for the biblical deluge in shaping current geologic features. These opinions elicited bitter opposition from conservative Christians on the Oxford faculty and throughout England. Yet Buckland remained deeply committed to the core principles of natural theology, which saw evidence of God's existence and beneficence in nature. "Minds which have been long accustomed to date the origin of the universe, as well as that of the human race, from an era of about six thousand years ago, receive reluctantly any information, which if true, demands some new modification of their present ideas of cosmogony," he explained in *Geology and Minerology Considered with Reference to Natural Theology*, "and, as in this respect, Geology has shared the fate of other infant sciences, in being for a while considered hostile to revealed religion; so like them, when fully understood, it will be found a potent and consistent auxiliary to it, exalting our conviction of the Power, and Wisdom, and Goodness of the Creator."

Buckland was a thoroughly rational Christian. When encountering an alleged miracle of martyr's blood perpetually wetting the floor of a Roman Catholic cathedral, for example, he tested the hypothesis by licking the spot with his tongue. "Bat urine," the self-satisfied Anglican cleric glibly pronounced. Buckland's God used systematic processes to guide terrestrial events with a designer's touch; his God did not intervene irrationally. For Buckland and many other early-nineteenth-century British naturalists, the succession of species in the fossil record reflected God's direction for life on earth. It had a beginning and human beings are its end. They never presumed to explain precisely how God created new species at the dawn of each epoch—as a divine act, that was beyond the realm of science. At most, Buckland would speak of divinely endowed laws of creation operating over time, but typically he offered no specifics of how they might work except to affirm that they could not involve evolution. Of course, for pre-Darwinian evolutionists, dinosaurs all but clinched the case of common descent. For both groups, notions of direction crept into biology and paleontology. Geologists trailed along.

In the same year that Buckland became Oxford's first geology teacher, 1818, Cambridge named Adam Sedgwick as its geology professor. He held the professorship for over fifty years, becoming an institution at Cambridge and within British science. A kindred spirit with Buckland in matters of science and religion, the pair set the norms for their field in England for a generation. Where Buckland did his best work on recent geologic features, however, Sedgwick sought to penetrate the fossil record back to the earliest vestiges of fossilized life on earth. This was virgin soil for geology in the 1820s, and Sedgwick found an ideal place to look for it in the ancient rocks of Wales. There Sedgwick "discovered" the Cambrian system (so named in 1835), the oldest strata of fossil-bearing rock, deep in the Transition series underlying Cuvier's Secondary formations. Here trilobites reigned.

Summer after summer, Sedgwick worked the ancient Welsh strata from the north while Sir Roderick Impey Murchison, a retired army officer who later led the British Geologic Survey, worked it from the south. They found fossilized fishes in the upper reaches of the Transition series but only the remains of long-lost invertebrates (like trilobites) in the Cambrian. The separate, well-defined age of invertebrates and age of fishes appeared to precede the familiar age of reptiles and age of mammals in the fossil record.

In 1841, John Phillips (nephew and student of pioneer geologic cartographer William Smith) formally divided the geologic column temporally by bestowing new names on its layers. The old Transition series became the Paleozoic Era, an age of invertebrates and fishes; the middle Secondary series became the Mesozoic Era, an age of reptiles; and the young ternary series became the Cenozoic Era, an age of mammals. These eras were defined by sharp breaks in the fossil record, as were the various periods within each era (like the Cambrian), but a trend line stood out. "Now I allow (as all geologist must) a *kind* of *progressive development.* For example, the first fish are below the reptiles; and the first reptiles older than man," Sedgwick wrote in a letter to Louis Agassiz in 1845. This directional finding simply reinforced Sedgwick's pious creationism, for he immediately added, "I say, we have successive forms of animal life adapted to successive conditions (so far, proving design), and not derived in natural succession in the ordinary way of generation" by transmutation or organic evolution. "How did they begin?" Sedgwick asked; he answered, "I reply, by a way . . . I call *creation.*"

Sedgwick, Buckland, and other naturalists of their ilk saw a pattern of successive creation in the fossil record, with God as its active source and a cooling earth as its mechanical regulator. Acting in an epoch-by-epoch, discontinuous manner, God lovingly designed new populations perfectly fitting the ever cooling, ever improving terrestrial climate while mercifully destroying the preceding populations when they no longer fit, leading to the creation of humans in what some saw as the biblically prophesied end times. Reviewing the succession of forms in the fossil record from his Cambrian invertebrates to "the recent appearance of man," Sedgwick thus affirmed in his Presidential Address to the Geological Society of London, "There has been a progressive development of organic structure subservient to the purposes of life."

ACTION AND REACTION

In the half century since Cuvier and Werner came on the scene in the late 1700s, geology and paleontology had utterly transformed Western conceptions of the earth's past and probable future. Notions of progress became firmly embedded in the scientific and religious mind. Biblical conceptions of creation morphed in response to fossil discoveries and geological theories without any apparent loss of faith. If anything, religion had become stronger, at least in Britain and probably in France. The secular Enlightenment of Buffon and Lamarck had been corralled. Science and religion appeared as bedfellows in the work of Buckland and Sedgwick, and as fellow travelers in that of Cuvier and Werner. Hutton carried his religion on his sleeve, and it spurred his science. Lyell was only a bit more discreet. In every case, while contributing to science, this work provided fodder for Christian natural theology. The likes of Robert Jameson, Edward Hitchcock, James Dwight Dana, Thomas Chalmers, Hugh Miller, William Paley, and countless nineteenth-century preachers and priests presented it to the public as evidence of God's existence and beneficence.

If ever there was a golden age of science and religion in Western Christendom, this was it. Neither side dominated or suppressed the other; both sides found inspiration from the relationship. For Christians, it required reinterpreting scripture to fit the advances of science. For scientists, it involved accepting the idea of divine design in nature. Perhaps the specter of naturalistic evolution, raised by Lamarck and others during the Enlightenment, helped. Conservative

Christians undoubtedly welcomed Cuvier, Buckland, Sedgwick, and Lyell as allies in their battle against Lamarckism, under the reasoning that the enemy of my enemy is my friend. But the ready use of the geologic theories and fossil discoveries in natural theology betrayed a deeper alliance. In the wake of Enlightenment secularism and the French Revolution, science and religion simply got along. And given the concurrent work of Michael Faraday and John Dalton in physics, it was not limited to geology and paleontology—though those became the textbook cases for later historians of science. Of course, it didn't hurt that so many of these scientists were theists who understood religious concerns and thought in Christian terms.

The golden age ended abruptly. As we discuss in the next two chapters, the revival of evolutionism during the mid-1800s complicated the relationship between science and religion. The theory of human evolution became a flashpoint that could not be reconciled with the Genesis account as easily as Cuvier's concepts of an ancient earth and geologic ages could. By the mid-twentieth century, under the banner of "creation science," a militant band of so-called young-earth creationists—or creationists that confined their understanding of the earth's past to fit within their literal reading of the biblical chronology—extended their attacks on Darwinism to include modern geology and paleontology as well. With this, hard-shell Fundamentalists rejected the gap and day-age understandings that had once defined their special creationist beliefs.

Even dinosaurs became a pawn in this moment of warfare between science and religion. Conventional biology textbooks and natural-history museums still used them to illustrate how evolution works, but, recognizing their appeal to kids, creation-science tracts and displays began presenting them as gentle giants that roamed Eden with Adam and perhaps still survived in the jungles of New Guinea or South America. The Institute for Creation Science sponsors tours to such lands in the hope that, by finding living ones, it could give credence to the Genesis account that every kind of land animal was specially created by God and named by Adam within the past ten thousand years. As for the fossil record and the earth's features that Cuvier, Werner, Lyell, and their successors attributed to epochs of geological action, current creationists see them as the residue of the Genesis flood. The very idea of progress gives way to belief in a fallen earth that is on a rapid road to Armageddon.

Meanwhile, science has moved on. Modern geology has evolved to include plate tectonics, with the earth's molten core carrying the continents around the globe like slow-moving jigsaw-puzzle pieces. Colliding plates explain explosive volcanoes and earthquakes better than Lyell ever could, while hot spots account for shield volcanoes and oceanic island chains in a reasonable fashion. God seems irrelevant to the process except perhaps as a distant and unseen creator of the physical laws that have played out over the past 4.5 billion years.

And proving the rule that nothing in science is ever wholly wrong (or right), so-called catastrophism in paleontology—long obscured by the rise of uniformitarianism in geology—has reemerged. The breaks that Cuvier, Buckland, Sedgwick, and other early-nineteenth-century directionalists detected in the fossil record, and that Lyell sought to explain away, really do exist, although not quite as sharply as once thought. Paleontologists now agree that the earth has been wracked periodically by mass extinctions, with the worst of them caused by collisions with asteroids that kicked up enough dust to darken the sun for years and cool the earth for decades. Other extinctions had different causes, some not yet known. Such events killed off most living species on at least five separate occasions during the past half billion years, leaving the field open for surviving ones to evolve and fill niches not before available to them. Dinosaurs went out with last big one, opening the way for mammalian life to blossom. In a scientific account that no major religion has yet sought to reconcile with its belief system, humans then emerged. By their actions, people are now causing a sixth mass extinction, many biologists fear, and while some Armageddon-minded Christian Fundamentalists seem not to care, many other religious believers (both in the East and the West) are rallying to protect the environment whether as God's creation or humankind's home. More on that appears in chapter 10 on religion and the environment. For now, it is enough for the historian between us to say that nineteenth-century paleontology and geology offer a glimpse of what once was and again could be in the ongoing relationship between science and religion. We turn the story over to our philosopher for a close look at the impact of Darwinism on the evolving interaction between these two chief ways of knowing.

CHAPTER FIVE

Darwinism and Belief

EVER since Charles Darwin published his landmark *On the Origin of Species* in 1859, no issue has fueled the science-and-religion debate more than his theory of evolution by natural selection. Indeed, that scientific theory and religious reactions to it have come to dominate and define the debate. Separately and together, we are called on to teach, write, or speak on this one issue more than all the other issues in science and religion combined. To give it due weight in this book, now that we have discussed astronomy, cosmology, physics, geology, and paleontology sequentially, and to allow both of us to speak on this central topic in our respective fields, we devote two chapters to it. First our philosopher will tackle it from his lens, taking a somewhat more philosophical approach to the nature of the relationship between evolutionary science and various world religions. Then, in chapter 6, our historian will chime in with a somewhat more historical narrative chronicling how the relationship unfolded in place and time. To avoid repetition and move the story forward, this chapter focuses on Darwinism and religion in general, while the next concentrates on the debate over the theory of human evolution in particular. The philosopher between us begins.

Everyone knows families where the children look and act nothing at all like their parents and where their attitudes and lifestyles are so different that it causes friends to wonder if the hospital made a mistake and sent home some babies with the wrong parents. Then, one day, perhaps when everyone is a bit tired and the light is none too good, a gesture, a glimpse of a profile, a tone of the voice—and, oh my goodness, the child is not just a chip off the old block, he or she is a veritable cord.

Christianity and Darwinism—meaning Charles Darwin's theory of evolution through natural selection—are a bit like that. And so are Judaism and Darwinism. Of course they are different. They could not be more so. Christianity and Judaism affirm that humans were created miraculously by a good God, who not only made the world for humans to live in but who continues to care for people right through their lives and beyond. Darwinism counters that humans are the current end product of a long line of lower organisms, right back to blobs and beyond. If the Genesis account shared by Jews and Christians says God made the first human from the dust of the ground, then people are modified mud; and if Darwinism says the unbroken laws of nature, forever cycling mindlessly, brought humanity into existence, then people are modified monkeys, with that simian line itself emerging from a mudlike ooze of organic molecules.

And yet, then suddenly the evolutionist working away in the lab or out in the field, perhaps collecting frogs or primroses or digging for fossils or counting fruit flies, stops suddenly, brought up short. There is a flash of something, and for those who had Christian or Jewish backgrounds, no matter how long ago or pushed down by newer memories, it all comes flooding back. Perhaps it is a flower's intricate design to capture insects, or it is the realization that it all started somewhere way back when and is incredibly beautiful and a bit frightening, or it is a sense that there is something about human nature that simply doesn't seem to be reducible to molecules or whatever. Through Darwinism the philosophical observer sees something older, something farther back in Western culture—something that might spring from a Jewish or Christian view of life.

Those who know a bit about the history of science should not really be surprised by this observation. They know that Darwinian evolutionary theory came from somewhere and that it didn't come from Japan or China or India or Africa or any other non-Western part of the world. It came from Europe, western Europe, and although countries like Russia and obviously the more recent United States contribute to the story, it was places like Britain, Germany, and France that gave birth to such thinking.

And it didn't happen twenty-five hundred years ago when the Greeks began to think critically or a bit later when the Romans conquered much of Europe. It happened when Christianity was part of the very fiber of the culture of those countries. It happened when to make sense of the world and its inhabitants one turned to the stories of the Bible and to the philosophical schemes of Catholic thinkers like Augustine and Aquinas—later Protestant reformers like Martin Luther and John Calvin—and found what guidance and understanding was to be taken from these sources. By "guidance and understanding" in such a context, we want to suggest that there are at least three big questions of life that can be asked and that Christianity and its parent Judaism try to answer. This in itself is significant because it has not been obvious to every intelligent person at every time and in every place that there are these questions, let alone answers to them. When addressing these questions, this chapter focuses on Christians because their beliefs most strongly influenced European thought when Darwinism arose during the 1800s, but the answers would be much the same for Jews of the period. For Islam, the answers have too little a tie to the origins of Darwinism to be of concern here, but as a theistic religion that posits the divine creation of humans, the questions are much the same and the answers are not too different.

THREE QUESTIONS ABOUT ORIGINS

First big question: "Where did everything come from?" Today, as in Darwin's day, this might strike most thinking Europeans and Americans as one of the obvious big questions of life. But many ancient Greek philosophers—Aristotle, for instance—never thought to ask it. To him, things just were. No one asks, "When did 2 + 2 first start equaling 4?" Or if people do, they show that they don't understand the meaning of mathematics. The equation 2 + 2 = 4 always was and always will be. That is how it goes in mathematics. And that is how it goes in the real world, in the eyes of Aristotle. Matter just is. Of course, it can be altered. But the basic stuff simply is no more a subject for inquiry about origins than are the truths of arithmetic and geometry. Many religions are much the same—Hinduism and Buddhism, to name two major examples, and also Confucianism, if that can be called a religion: the universe just is. And people just are.

Yet within Christianity, the question about origins is one that is asked, and it is one that is thought coherent and worthy of an answer. It is clear why: Christianity is based on Jewish thought, and Judaism is imbued with a historical consciousness. The ancient Hebrew people did think in terms of origins and beginnings, and Christians followed them in doing so. This is what Genesis is all about. God made the world and set it in motion. That is the answer to where things came from "in the beginning." Believers needed to know that is true. It is central to their theology.

Second big question: "What kind of world do people live in?" Some people may have so little imagination that they would never ask a question like this. They just take things as they are, one event after another. But most people have some curiosity, and of course there are all sorts of answers one might give, at different levels. One of us—the historian—after living in Seattle for two decades, only half jokingly complained that it's "a world where it never stops raining!" while both of us have felt much the same after "summering" in England. Louis Armstrong can croon that it's "a wonderful world," while ancient Babylonians typically saw it as a dreadful place dominated by gods who were at best indifferent to humans and at worst malicious. Others might use words like "mystifying," "awe-inspiring," or "comforting." Responses like these get closer to some of the kinds of answers one finds in much modern religious thought, from mainline Protestantism through Reform Judaism to New Age mysticism.

More traditional Christians, those informed by natural theology, might go in a slightly different direction and say something like, "We live in a world that is put together in a pretty clever fashion. It isn't just a random heap of bits and pieces. The rains fall, the crops grow, the animals and humans have food to eat, and so life goes on in an orderly, functional fashion." This is a very Christian sort of answer, and it is one that finds support in the Bible. It was also very much the way of Plato and Aristotle. It was Plato who drew the inference that such a beautifully functioning world could not be pure chance, there must be an intelligence (aka God) that lies behind it. Aristotle was more biologically minded than Plato. It was he who saw that it is in the living world of plants and animals that one really finds this intricate functioning, this designlike nature, which seems to be more than just random chance. Living organisms are such clever structures that there must be some meaning to them. And it is here that Christian theologian-philosophers, from Thomas Aquinas to William Paley, stepped in to make sense of it all. To them, the world does seem like it is designed, because it *is* designed—by an all-powerful, all-loving being—the God of Christianity.

Third big question: "Where do humans fit into the scheme of things?" At least by their way of judging, by any relative measure, humans are way ahead of all other organisms, even ahead of—especially ahead of—those beings that seem most like them, the higher apes—gorillas, chimpanzees, orangutans. Humans may not be as fast as the cheetah or able to fly like the butterfly. They may not be as beautiful as a flower in the morning. But they have intelligence and the tools—in particular arms and hands, but also an upright nature—to do something with them. Humans also have a consciousness of their own existence and of a sense of morality and mortality far beyond that of any other kind of animal. They are not gods, but it is not surprising that the world's most widely accepted religious account of human origins maintains that people are made in the image of God and are given "dominion" over the earth and all its nonhuman inhabitants.

THE ADVENT OF EVOLUTIONARY THINKING

Now where does the theory of evolution—Darwinism in particular—fit into all of this? Well, to be quite frank, nowhere very much until after the so-called Scientific Revolution, which is discussed in earlier chapters. Nicolaus Copernicus showed that the earth goes around the sun rather than vice versa, and following this, astronomers like Johannes Kepler and Galileo Galilei mapped the heavens and discovered the laws that govern terrestrial motion. Finally, Isaac Newton tied everything together with his law of universal gravitation.

Then people started to speculate in a naturalistic way about physical origins: Did the planets always circulate endlessly around the sun or was there a time when this all started to happen naturally? As noted in chapter 1, the late-eighteenth and early-nineteenth-century French scholar Pierre-Simon Laplace suggested that the solar system had evolved over time naturally from a rotating gas nebula. And as discussed in chapter 4, geology too was playing its part through an increasing sense of directional change over time, as described in the work of such influential early-nineteenth-century naturalists as Abraham Werner and Georges Cuvier. With the advent of science came technology and the development of new instruments and machines—the Newcomen engine, for example, a pump for removing the water from mines. Mines were increasing in number as people looked for needed minerals and above all for coal, the new and more powerful fuel taking over from the traditional wood. All of this focused people's thinking on the earth and its nature and whether it is stable or in some state of constant evolution and change. Did minerals grow in the earth? some asked, and Where does coal come from?

And then there was the organic world. The focus on design implied stability. For example, the human or animal eye is just so perfect, it must have always been so, like a cleverly designed watch or, to use the eighteenth-century British natural theologian William Paley's analogy for it, a telescope. But there were pointers the other way. Perhaps most significant was an idea that went back to the medieval era, and through that, even farther back to some Greek philosophers. Could it not be that the whole of life might be placed in an upward pointing "Chain of Being"? In it, the most primitive organisms are at the bottom and then, climbing up through the fish and reptiles and mammals, the humans are reached at the end. This is not a dynamic picture—it is a ladder fixed in eternity rather than an escalator moving in time—but obviously it does lend itself to speculations, and if change is in the air, so be it. For the late-eighteenth-century French evolutionist, it became an escalator. Backing this temporalizing of the chain were increasing numbers of embryological studies, where the sequence from the fertilized egg to the grown adult can be followed in detail—again giving rise to speculations about change in the broader context. Then, with more mining and the cutting of canals across the European countryside, the unearthing of fossils gave direction to animal development the deeper people dug, from more complex above to simpler below—a seeming temporal record of progressive change over time documented by Cuvier, William Smith, and others discussed in chapter 4.

Above all, however, what truly triggered naturalistic speculations about organic origins was the new philosophy of progress. This is the idea that change for the better—in society, in well-being, in medicine, in science, in culture generally—can be achieved through human abilities and effort. Humans unaided except by their natural or endowed reason can make things better. This was not really an idea to be found in ancient Greek philosophy nor indeed in the Christian West until the Scientific Revolution had given birth to the Enlightenment, when all things starting looking brighter on this side of eternity. By the eighteenth century, however, people were starting to think this possible—a better life here and now—and increasingly there were voices speaking in its favor. It was, for instance, a powerful idea in Britain's North American colonies and undoubtedly played a major role in motivating the American Revolution. Pertinent to this story is that as soon as people started speculating about human progress, they started looking for confirmation, and the natural world of animals and plants soon caught their attention. If organisms can indeed be put in a chain, why not make it dynamic, with those at the bottom over time climbing upward and eventually being transformed into the highest form of being—humans. Take note: no one had much evidence. Back at the beginning of the eighteenth century, there was not yet a detailed fossil record.

But for all this, gradually people were starting to think in evolutionary terms—although back then the word "evolution" referred more to embryological growth, and people tended to use such terms as "development" and "transmutation" for biological change over time, as in the transmutation of one species into another. So much for God specially creating each biologic kind at the beginning. Species might develop much like individual embryos—so much the grander and still potentially divine. A wholesale God beats a retail God, in the progressive mind. Even if humans were made in God's image, people began to think that evolution seemed more apt than stasis, though many who thought this way did not give much of an ongoing role to God beyond the creation. Erasmus Darwin, the grandfather of Charles, in his "Temple of Nature" (1802) put this all in verse:

> Organic Life beneath the shoreless waves
> Was born and nurs'd in Ocean's pearly caves;
> First forms minute, unseen by spheric glass,
> Move on the mud, or pierce the watery mass;
> These, as successive generations bloom,
> New powers acquire, and larger limbs assume;
> Whence countless groups of vegetation spring,
> And breathing realms of fin, and feet, and wing.

DIFFERENCES AND SIMILARITIES

Consider this hypothesizing in context, in particular of the context of the then dominant religion of western Europe and the Americas—Protestant Christianity. As it happens, few of the early evolutionists, if they may be called that without undue anachronism, were atheists or even agnostics. Like Erasmus Darwin and his science-minded friends, such as Benjamin Franklin and Joseph Priestley, they might have trouble with the divinity of Christ. But they all thought him a wonderful role model, and behind him they all accepted the existence of God—a God who had created and set the world in motion and still watched over it, and us, with a beneficent eye—the God of deism or theistic Unitarianism. Nevertheless, they knew that they were going against central aspects of Christianity. At the time, they and their Christian critics were generally not greatly worried about literal aspects of the Bible—obviously an evolutionist cannot take the first chapters of Genesis as literally true, word for word, but from Augustine on, sophisticated Christians had been realizing that aspects of the Bible must be understood metaphorically, typologically, or allegorically. But those Christian critics were worried about progress, for this stood in direct opposition to what they took to be central, Providence, which is the idea that sinners can do nothing to save themselves other than rely on faith in God's forgiveness.

Grant these differences. But what is fascinating are the similarities, most particularly the similarity that the evolutionists took so seriously: the three main questions that were of such concern to Christians. Above all, Darwinism is a story of origins. No nonsense about eternal existence or whatever. It is telling about where everything came from, organic species in particular. Today, scientists tend to separate cosmogony, the origins of the universe, from evolution, the origins of species. Back then, however, it tended all to be part of one story. And this package deal may not have been the Christian answer, but it was an answer to the same question posed by the Christians: Where did everything come from?

It is similar to the second question, about the nature of the world. For Christians, believing in a good God who was all-powerful, the world (including the world of organisms) simply had to be well designed. God would not have done—could not have done—a botched-up job. The eye does not function as well as it does just by chance. Granted, this is a bigger question for Western Christians—Catholics and Protestants—than Eastern Christians because the former focuses more on this world than the latter. But it should be remembered that one of the leading evolutionary biologists of the last century, Theodosius Dobzhansky, was a lifelong Russian Orthodox Christian, and he too mulled over this question. It is relevant, however, that he was the product of robust schools of modern genetics and field naturalism that arose in the late czarist period and flourished under the early Soviet state. Both schools clearly benefited from an increase in Western secularism during the former period, and establishment of official atheism under the latter state as counterbalances to the Orthodox religiosity of Old Russia. Most of Dobzhansky's teachers and colleagues prior to his emigration to America, and many of his colleagues and students in the United States thereafter, were agnostics or atheists. Perhaps reflecting his years in the Soviet Union, some of these American collaborators describe him as someone who compartmentalized his science and religion in such a way that neither impacted the other. But back to the Western Christians for whom this question loomed large.

To be candid, before Darwin, early evolutionists did not have a ready answer to the question of design in nature. Their Christian critics said that blind law does not lead to complex functioning—it leads to rust and decay and disorder and things not working—and generally the evolutionists thought they had a good point. The eye is pretty convincing. How could something so functionally complex come about by chance? Early evolutionists pretty much had to tough this one out, hoping that a solution would emerge. As it did, in 1859 when Charles Darwin published his *On the Origin of Species.* Just as Newton had put forward a mechanism to explain the Copernican universe—the force of gravitational attraction—so Darwin put forward a mechanism to explain or account for organic evolution—natural selection. Darwin seized on the calculations of Thomas Robert Malthus, a turn-of-the-century (eighteenth to nineteenth) English cleric who obsessed about overpopulation. The rate of reproduction always outstrips the available food and space. This means that there will be a fight for resources, what Malthus called a struggle for existence. He was thinking about people, but the same reasoning would apply to competition among organisms of any given type or species.

Darwin, who was a terrific naturalist, knew that given any population of organisms there is always going to be a huge amount of variation. Such differences can impact success in a

Malthusian struggle for existence. If food supplies are short, then perhaps the smallest of any given type will survive—and more importantly reproduce—and the largest will perish for want of sustenance. Again, if predators are present, even the slightest variation in speed or coloration or strength might make the difference between living long enough to reproduce and dying without offspring.

In other words, all of the time there is going to be a natural winnowing of organisms, with (on average) those with locally beneficial variations—what Darwin called "adaptations"—being the winners. This process, using an analogy from the practices of animal and plant breeders, Darwin called "natural selection." The important thing is that this doesn't just lead to change, or evolution, it leads to change in a particular direction, toward ever more beneficial or efficient adaptations. More-efficient eyes help their possessors in the struggle for existence—the same for grasping hands, sharp teeth, and everything else. In other words, Darwin gave an answer to the second big Christian question: What kind of world do we live in? Why is the world as well designed as it is? It was not the direct intervention of a good God but the end result of the slow, law-bound process of natural selection.

What about the third question, where human beings fit into the scheme of things. Let there be no doubt that the early evolutionists were as obsessed by the superior status of humans as anyone else. Progress leads up to human beings, they assumed. But Darwin made things a bit more complicated. Socially, he was as much in favor of progress as anyone. His maternal grandfather was Josiah Wedgwood, the commercial potter, one of the leaders of the Industrial Revolution.

Darwin was not about to turn his back on upward change. Biologically, he had little doubt that humans came out on top. The trouble is that natural selection seems to be relativistic. It is not a tautology—tautologies are things that are necessary and the struggle for existence is not that, for it might never have happened—but it does suggest that winning and losing is not an absolute thing but depends on the circumstances. Little food and it pays to be small. A lot of food and being big might be the way to go. And foodwise, humans are pretty high-maintenance organisms. Compared to some other species, humans are not very strong or fast, so they have to make their way by cooperation, intelligence, and tool use. That means big brains, or at least that was how Darwin reasoned.

Obviously, if the theory of evolution is to be believed, it worked. Otherwise humans would not be here. But it was not obviously going to succeed. Darwin solved the problem by appealing to what evolutionists today call "arms races." Lines of organisms compete against each other and adaptations improve. The prey gets faster and so the predator gets faster. Eventually intelligence emerges as it proves its worth in the races, and humans emerged on top. Darwinism, like Christianity, supplied answers to three fundamental questions that Westerners cared about, which made it appealing to many of them.

DARWINISM AND BUDDHISM

Having said all this about Western Christianity (and by implication the major Western theistic religions), what about the Eastern religions—how do they relate to Darwinism? Take Buddhism as an example, but much of this also applies to Hinduism and some other Eastern religions. Instinctively, many people assume that Buddhism must be compatible with Darwinism because there is no long history of antievolutionism in Buddhist countries. And perhaps there is some truth in all of this, but that is not quite the issue here. The issue here is how Christianity as a world system measures up against evolutionary theory as a world system, and the similarities might inflame the conflict. So, how does Buddhism as a world system measure up against evolutionary theory?

Just as there are different versions of Christianity, so there are different versions of Buddhism. But, as outlined more fully in chapter 3, the basics are fairly straightforward. Although it dates back to one individual, the Buddha who lived about five hundred years before Christ, he was not a god or the son of God but rather one who was particularly enlightened. According to his teaching and that of his followers, at a temporal level there is no beginning or end—all is eternal. People are born into the existing world, but when they die that is not the end, for they will be reincarnated, indefinitely, until they achieve a blessed state, *nibbana* (nirvana), when they get off the Chain of Being. Spatially, there are different levels of existence, with humans about in the middle. Right at the bottom is hell, *niraya*, with vile beings tortured. Then the scale moves up through animals to humans, and then on to the levels of gods—the lesser gods, the *asuras*, and then the higher gods, the *devas* including the *brahmas*, the very highest form of godlike being. All are subject to the same laws, and through reincarnation all move up (or down) the scale, with the goal of finally getting off entirely.

Return then to the three questions that linked Christianity and Darwinism and note the lack of correspondence with Buddhism. Like the Greeks, the Buddhist does not even try to answer the first question, about where people and everything else come from. Existence is eternal—always was and always will be. It changes, of course, for beings are born, live, die, and then are reincarnated. But as with the Greeks it is change within limits. There are gods—it is a mistake to say simply that Buddhism is atheistic. But there is no Creator God, responsible for everything. It is like the 2 + 2 = 4 situation. No beginning, no middle, no end.

The second question goes much the same way. Of course, a Buddhist can explore the world and its nature. In fact, this is part of the duty laid on humans. But it is interesting how little this exploration is about God. With no Creator God there is no question of this Creator being good or bad, powerful or helpless, making an intricately functioning world or not. The Buddhist is not about to deny that the eye exists and that it works well, but it is not evidence of anything, nor does it demand special explanation. The world exists as it is, that is all and that is enough.

Finally, about the status of humans. While humans are not at the bottom of the heap, they are not the top of the heap either. There are the *asuras* and *devas* above them. And it cannot be overemphasized that these different levels of existence are not connected to the Creator God. There is no such being.

At the philosophical level, then, the theory of evolution and Buddhism will not conflict, cannot conflict. If a Buddhist wants to integrate evolution into the world picture, there is nothing stopping this. In fact, Buddhism emphasizes the immensity of time. There are multiple universes, and what happens in any one goes on indefinitely. To use one Buddhist metaphor, thinking of a universe, one thinks in terms of eons, and a cloth stroking a mountain and reducing its size would be finished with its task before an eon was finished. If the evolutionist wants to explain the eye in terms of natural selection, then so be it. That is not the topic that interests or threatens the Buddhist as Buddhist. Even the place of humans—perhaps natural selection did produce them. Given the infinity of time, who knows what has happened? In any case, moving up or down the levels of existence is a matter of moral behavior and has nothing to do with physical causes.

THE PROBLEM OF EVIL

Historically, one of the biggest problems facing the Christian is the problem of evil. How could a God who is all-powerful and all-loving permit pain and suffering? Why do bad things happen to good people? Expectedly, the Christian has a number of answers to these issues. Traditionally, one separates evil into two kinds. First there is moral evil brought on by human beings. They are made in the image of God but live in a fallen world of their own sinful choices. It is better—it is more human—to have free will and then choose Christ than to be bound, to be determined. This is so, even though it can lead to dreadful consequences. Second, there is natural evil, like the mutation causing Huntington's chorea or Tay-Sachs disease. Here the usual defense is a version of the argument made famous by the seventeenth-century German philosopher Gottfried Wilhelm von Leibniz. He pointed out that God's being all-powerful does not mean God can do the impossible. God cannot make 2 + 2 = 5. Making the world was a matter of balances. If God was going to warn us against the fire, then God needed a pretty powerful method of motivation—pain. On balance, the pain from burning for some outweighs the dangers from burning for all. Likewise with mutations. If God was going to have some way of making organisms, including humans, then in the end God needed mutations. On balance, the good ones leading to such things as intelligence outweigh the evils of the mutations that lead to pain and suffering. And how much can any suffering in this brief life matter anyway when balanced against eternity? All things lead to good in those that trust in God, Christians can affirm.

This is not a book on the philosophy of religion, so we need not speak to the adequacy of these defenses. We do want to do two (related) things. First, we point out that the problem of evil is a Christian issue. It stems from the kind of God the Christian supposes. That is, an all-powerful, all-loving being. Just because something is a problem for Christians does not necessarily imply that it is going to be a problem for other religions. This is so of evil. For Buddhists, for example, it is virtually irrelevant. This is not to say that the Buddhist is insensitive to human suffering. Far from it. The whole point of the system is to acknowledge it and try to deal with it. The aim is to escape from *dukkha*, suffering, and to achieve *nibbana*, release. One does this through proper living. But evil is not a *problem* for the Buddhist in the way that it is for the Christian. There is no one at the back of things to blame for making the world the way that it is. The world just is, and evil is part of it.

Which brings us to the second point about the problem of evil. It may not be such a big problem for non-Christian religions, but what about under evolutionary biology, especially the Darwinian version? If Darwinism is so close to Christianity, is the problem of evil something with which it wrestles? Now, obviously in one sense it doesn't and can't. Science doesn't work like that. Science seeks to tell it like it is or appears to be based on testable observations and repeatable experiments.

Which, of course, is true, but it doesn't mean that science cannot have something of interest, of pertinence, to say about the problem of evil. After all, Darwinian theory came into being creating and defining itself in relation to Christianity, and it attempts (perhaps because of or perhaps despite of) to answer the kinds of questions Christians find pertinent. As it happens, this suspicion is borne out. In particular, the strife and suffering caused by the evolutionary process has been taken by some as yet one more nail in the Christian coffin, yet one more reason why the problem of evil is simply an insuperable objection to the Christian's claims or at least to Christian natural theology. Charles Darwin himself felt this, writing to his good friend the Harvard botanist Asa Gray, who was incidentally a sincere Presbyterian. "I had no intention to write atheistically. But I own that I cannot see, as plainly as others do, & as I should wish to do, evidence of design & beneficence on all sides of us. There seems to me too much misery in the world. I cannot persuade myself that a beneficent & omnipotent God would have designedly created the Ichneumonidae with the express intention of their feeding within the living bodies of caterpillars, or that a cat should play with mice." Yet isn't that all logical under natural selection—the problem of evil resolved from a Darwinian perspective as part of the evolutionary process? A Calvinist Christian, Gray understood this perspective—only a few are selected, whether by God or nature—so it did not pose an insurmountable problem to his religious faith, but natural selection was identified early on as an issue by some Christian critics whose faith rested heavily on natural theology, such as the influential Princeton theologian Charles Hodge. "What is Darwinism?" Hodge asked in an 1874 book with that title. "It is atheism," he answered, because the exclusion of divine design from nature "is tantamount to atheism."

Darwin in fact pulled back somewhat, saying that although he could not see evidence of a friendly God, he didn't want to say there isn't one. But others, less charitable to Christianity, have taken up this point and run with it. Richard Dawkins, science writer and prominent modern atheist, has made much of it. He points out that the adaptations of the cheetah are for killing gazelles, and the adaptations of gazelles are for getting away and thus starving cheetahs. Obviously, the God of the Christians does not exist or he wouldn't have ordered the world in this way. To which the Christian can respond, How are they to know the mind of God and what such interactions mean in a fallen world? Can the crafted pot know the potter's purposes?—like the biblical prophet Abraham as present in the Christian existentialist Søren Kierkegaard's 1843 book.

FEAR AND TREMBLING—MUSTN'T THEY SIMPLY TRUST?

This suggests why science and religion can get entangled. An awful lot depends on what happened in the past, which evolutionists knew all along. If science and religion come from different stables, like Darwinism and Buddhism, then why would anyone expect them to interact, positively or negatively? But if they come from the same stable, like Darwinism and Christianity, then expect interaction.

So accept that science and religion can interact. Does this mean that they are always going to interact negatively? The example of the problem of evil suggests that this will be so. Indeed, much in this chapter points that way. It has not made too much of things like the six days of creation or the flood of Noah's time, but with the coming of Darwinism as well as modern astronomy, anthropology, geology, and so forth, these biblical accounts don't stand up too well as literal events. Then there is the clash between Providence and progress—about as big a difference as can be imagined—and so the story goes.

Actually the story is a little more complex and interesting than this. Obviously the development of science, not just evolutionary biology, may well challenge long-held (Christian) religious accounts. Six days of creation and the great flood among them. But this is nothing new. Augustine, back around 400 CE, faced up to them. He argued that the ancient Jews did not know science, so why should God speak to them in that language rather than in typologies, metaphors, and the like? It doesn't mean that he threw everything out—indeed, he kept everything that most Catholics and Protestants hold essential: God is still the Creator, humans are still made in God's image, and faith in Christ is still the means to salvation—but the quasi-scientific gloss must go.

As far as issues like progress and Providence are concerned, this is complex, admittedly, but start with the fact that a lot of people—a lot of non-Christian people—are today a lot less enthused about prospects for progress than they were back in the nineteenth century. After a century that saw the slaughter during the two world wars, not to mention the Holocaust and other genocides, plus mass slaughter in Russia, China, Cambodia, Uganda, and elsewhere, it is not easy to speak confidently of inevitable progress. But even if there is progress, for the Christian this does not mean that God's aid is necessarily unneeded. And after all, for the Christian, if humans are improving things, they are using the talents given to them by God, which is hardly something that goes against the messages of the Gospels. Christians can pray for healing even as they pray for God to guide their surgeon's hand and thank modern medicine for seemingly miraculous new drugs and medical procedures. Both can be seen as gifts from God—either as special or common grace.

What about evil? Even if Darwinism is relevant to the discussion, it is not introducing a new factor. People knew about pain and suffering long before the Anglican divine Malthus wrote on the struggle for existence and Darwin took it up and made it central to his theory. And if that is how God chose to create species by natural law, a Christian might say, how can humans (the pottery) presume to question the potter? The pain and suffering may simply be part of the package deal producing serviceable species, including *Homo sapiens.* Which all goes to show that, instead of being inevitable foes, Darwinism and Christianity might in some respects be compatible—or that at least was what the Christian Darwinist Asa Gray thought and said in his theory of theistic evolution, which still influences such evangelical Christian biologists as American geneticist and National Institutes of Health director Francis Collins. The Scottish New Testament scholar and retired Anglican bishop N. T. Wright said much the same about evolution as God's means of creation. To go back to the metaphor at the beginning of this chapter, family relationships can be very tense at times and yet somehow work out in the end. Now we will give our historian a chance to carry this matter forward from his discipline's perspective.

CHAPTER SIX

The Evolution of Humanity

THE last chapter discusses various religious responses to the theory of evolution, some of them hostile, others accepting. Speaking from a historian's perspective, however, the big issue has never been the theory of evolution in general, but applying it to humans. After all, many people care more about humans than they do about other animals. And who cares if plants evolved? But many people find the idea of descending from monkeys or being related to apes as really quite degrading to their self-image. For some, it is a religious matter, especially if they take the passages about Adam and Eve in the Bible or the Qur'an literally. But even if they don't go that far, theists who believe they were created in the image of God—and see that as what makes humans special—can become really quite testy about the theory of human evolution, particularly if it extends to include the vaunted human brain, emotions, and morality. This is nothing new.

An 1871 cartoon in the British magazine *Punch* shows an earnest young husband reading to his wife and infant child from Darwin's just-published book *The Descent of Man*. "So you see, Mary, baby is descended from a hairy quadruped, with pointed ears and a tail," he explained. "We all are." The wife responded tartly, "Speak for yourself, Jack. I'm not descended from anything of the kind, I beg to say; and baby takes after me." In saying so, she spoke for many. Although countless Victorian-era Europeans and Americans accepted the theory of evolution as applied generally to other species, and even perhaps the evolution of the human body, most of them refused to believe their minds, morals, or emotions evolved from those of beasts.

The centrality of the theory of human evolution to the modern science-religion dialogue is illustrated by the rise of Protestant Fundamentalism in the United States and Canada during the early twentieth century. At the time, responding to developments in science and the seeming progress of Western civilization, mainline Protestant theologians increasingly accepted an evolutionary view of human origins and adopted an allegorical approach to the Genesis account of creation and other scriptural passages, leaving parishioners adrift in the shifting currents of biblical interpretation. Some ministers and a growing number of faithful churchgoers reacted by defending the so-called fundamentals of biblical Christianity, with those fundamentals inevitably including a historical Adam and Eve, with Adam's fall seen as introducing evil into God's perfect creation and necessitating a second

Adam—Jesus—to atone for humankind's original and ongoing sin. For many, such a view made sense of the dark side of modernity while giving hope for eternal life in heaven.

At first, American Fundamentalists railed mainly against Darwinism in seminaries and churches. By the 1920s, however, they began crusading against teaching evolution in public schools. A hodgepodge of state and local restrictions resulted, leading to the sensational trial in 1925 of John Scopes for violating Tennessee's so-called Monkey Law. Tellingly, these early laws only barred teaching human evolution, perhaps because that issue generated the widest popular support.

The crusade's self-appointed leader, the silver-tongued politician and former secretary of state William Jennings Bryan, a Presbyterian lay officer, publicly expressed his willingness to accept cosmic evolution, earlier geologic epochs, and the evolution of other species, if proven, but drew the line on the special creation of Adam and Eve in God's image. To this doctrine, he would hold fast on biblical authority against all the claims of science. Teach children that they descended from monkeys, Bryan charged, and they would act like apes. What else could explain the spread of cutthroat capitalism, labor militancy, global imperialism, atheistic communism, and mechanized warfare? The scientists who developed the chemicals and other weapons that killed untold thousands during World War I believed in Darwinism, Bryan claimed, and the two went hand in hand. By World War II, even as mild-mannered an evangelical as Oxbridge scholar C. S. Lewis agreed that, while he accepted the theory of evolution, he too drew the line when Darwin began monkeying with man.

In some ways, little has changed. Many religiously inclined evolutionists still draw the line on materialism when it comes to the ascent of the human spirit or soul. In *Evolutionary Theory and Christian Belief*, Oxford University ornithologist David Lack, an Anglican theist whose study of Darwin's finches gave wing to the modern neo-Darwinian synthesis, wrote, "Science has not accounted for morality, truth, beauty, individual responsibility or self-awareness, and many people hold that, from its nature, it can never do so." In his 2002 article "Human Genetics," American geneticist Francis Collins, a self-described evangelical Protestant who directed the Human Genome Project and heads the National Institutes of Health, added, "Science will certainly not shed any light on what it means to love someone, what it means to have a spiritual dimension to our existence, nor will it tell us much about the character of God." Collins calls himself a theistic evolutionist.

Opinion surveys suggest that Collins speaks for about two-fifths of American people, and a similar fraction of American scientists, when he posits God as somehow involved in the evolutionary process—at least to the extent that God breathed a supernaturally created soul into a naturally evolved body to start the human race. Another 40 percent or more of Americans, those surveys report, see God as directly creating the first humans. Surveys taken in Muslim countries find even higher support for creationism. The Roman Catholic Church accepts either approach to human origins, but not Darwinian materialism. "Rather than *the* theory of evolution, we should speak of *several* theories of evolution . . . materialistic, reductionist, and spiritualist," Pope John Paul II wrote in a 1996 message to the Pontifical Academy of Sciences, reiterating that believing in theistic evolution was okay for Catholics. But he promptly added, "Theories of evolution which, in accordance with the philosophies inspiring them, consider the [human] mind as emerging from the forces of living matter, or as a mere epiphenomenon of this matter, are incompatible with the truth about man." Most of the world's leading biologists would surely disagree.

IN THE BEGINNING, DARWIN

From the earliest days of Darwinism, what people found most intriguing and discomforting was the idea that humans' higher attributes were not intelligently designed by God but instead obtained through a process of natural selection. "I have read your book with more pain than pleasure," Cambridge University's devout Anglican geologist Adam Sedgwick wrote to Darwin within a week of receiving a prepublication copy of his former student's *On the Origin of Species*. "Tis the crown and glory of organic science that it *does* thro' *final cause*, link material to moral. . . . You have ignored this link; and, if I do not mistake your meaning, you have done your best in one or two pregnant cases to break it." For Sedgwick, that final cause was creating humans to glorify God. Although he closed his letter to Darwin with a lighthearted reference to himself as "a son of a monkey," Sedgwick never gave ground on a divine source for the human soul. For him and many other Victorians, that was no laughing matter.

The idea of human evolution, although linked most often to Darwin, was something he only publicly committed himself to gradually. Not wanting to put off readers, Darwin barely mentions the issue in *On the Origin of Species*, and he sticks to the origin of seemingly lower species. He had humans in mind all along, however, and came to see them as situated firmly in a progressive chain of being arising from below. Some of the earliest entries in his scientific notebooks on evolution draw really quite graphic comparisons between the native peoples of Tierra del Fuego, whom he met during his Beagle voyage in 1832 and considered the lowest form of humanity, and the primates in the London zoo. "Not understanding language of Fuegian[s], puts [them] on par with Monkeys," he wrote in 1838. In a later entry, he demanded, "Compare, the Fuegian and Orangutan, and dare to say difference so great."

Frequently, Darwin would speculate in these private notebooks about animal origins of human traits, writing, "One's tendency to kiss, and almost to bite, that which one sexually loves is probably . . . due to our distant ancestors have been like *dogs* to bitches." As for the vaunted "mind of man," Darwin concluded, it "is no more perfect than instincts of animals." Human thought itself, like animal instincts, he attributed to brain structure—and then chided himself "Oh you Materialist!" for thinking so. But none of this appeared in *On the Origin of Species.*

Yet as quoted in the preceding chapter, when Darwin's longtime American correspondent, the noted Harvard botanist Asa Gray, received his advance copy of *On the Origin of Species* and expressed to Darwin concern about its theological implications, Darwin replied that while he hadn't intended to write atheistically, he could not believe in the divine design of a natural world so filled with suffering, misery, and apparent cruelty as ours. Then, alluding to Protestant theologian

William Paley's famous analogy between a crafted watch and the human eye, Darwin noted in his letter to Gray, "Not believing this, I see no necessity in the belief that the eye was expressly designed." Even human nature and mental ability might result from natural processes, he added.

The sequence in Darwin's letter to Gray is telling from a Western religious perspective. It passed quickly from observations of what seems evil in nature (such as cruel animal behavior) to their implications for what seems well-designed good in it (such as the human eye), and then moved on to ponder the origin of what seems positively good (such as human morality and mentality). Few Christians want to blame God for the first; many could go either way on God's role in the second; but all want to attribute the third to their God. Without saying so, Darwin is leading Gray along with a carefully contrived theological argument. But Gray did not buy it. A gifted naturalist with unshakable Christian convictions, Gray certainly got the point—and though he came to accept the theory of evolution, he always maintained that God guided the process.

THE DESCENT OF MAN

After *On the Origin of Species* was published, Darwin avoided publicly commenting on human evolution for over a decade and left that matter to such followers as T. H. Huxley and Ernst Haeckel—both ardent secularists. It was not until *The Descent of Man* in 1871 that Darwin laid out his case that humans evolved along with other primates. Although intelligently designed, the argument was neither as compelling nor as widely accepted as the one in *On the Origin of Species.* Nevertheless, it has held up surprisingly well.

First, Darwin presented the evidence—well known by then—for the evolution of the human body. In anatomical structure and embryonic development, people resemble other animals, he noted, and the persistence of monkeylike rudimentary features such as the tailbone reinforces the conclusion that the human body evolved from lower forms. Relying primarily on structural similarities, Darwin traced human ancestry from "the most ancient progenitors in the kingdom of the Vertebrata," through ancient fishes and amphibians, early marsupials and placental animals, to "the New World and Old World monkeys; and from the latter, at a remote period, [to] Man, the wonder and glory of the Universe."

In a second volume—necessary, as he explained, for treating "the whole subject in full detail"—Darwin elaborated on the theory of sexual selection, which he had introduced in *On the Origin of Species.* Sexual selection, he believed, played an important part in external racial differences and accounted for the evolution of male mating traits, such as the peacock's tail. Applying this theory to humans, Darwin observed that peoples differ in what they find attractive in mates. Africans prefer dark skin and depressed noses, while Europeans favor light skin and straight noses, he wrote, while "in Java, a yellow, not a white girl, is considered . . . a beauty." Within each race, as the most attractive mates are chosen first and have the most children, sexual selection propagates favored external characteristics. For instance, Darwin claimed that Hottentot women have large bottoms because tribesmen desire mates who display that trait.

The body's evolution, even if accepted, did not settle the matter. At the time, many Christians believed that humans stood apart from animals due to their souls, not their bodies. So Darwin extended his naturalistic analysis to include those mental and moral attributes that supposedly uplifted humanity: higher reasoning, self-consciousness, religious devotion, and the ability to love. He asserted that humans' mental powers and moral feelings differed in degree, rather than in kind, from those of other animals, with a progressive gradient linking the lowest beasts to the highest humans, and he stressed the humanlike qualities of higher animals (particularly pet dogs and wild monkeys) and the animal-like qualities of the "lowest" savages.

"Can we feel sure that an old dog with an excellent memory . . . never reflects on his past pleasures in the chase? And this be a form of self-consciousness," the dog-loving Darwin wrote in a typical passage. "On the other hand . . . how little can the hard-working wife of a degraded Australian savage . . . reflect on the nature of her own existence!" Similarly, Darwin doubted whether Fuegians felt religious devotion, yet saw "some distinct approach to this state of mind in the deep love of a dog for his master."

Could distinguishing human characteristics—like abstract thought and a sense of morality—have evolved by a naturalistic process, Victorian evolutionists asked, or did God implant them in an evolved human body? Christian theologians had credited an indwelling soul, the existence of which lifted humans above other animals. Scientists generally segregated humans from other animals on this basis as well—from Aristotle's theory of the rational soul found only in humans, to

Cartesian dualism splitting physical matter from the human and divine soul, to Georges Cuvier's division of humans and primates into separate taxonomic orders. Now humans were lumped in the same order with other primates by Huxley.

Darwin attributed the evolution of even the most ennobling of human traits to gradual, survival-of-the-fittest processes. Long ago in Africa, he suggested, some anthropoidal apes descended from the trees, started walking erect in the open spaces, began using their hands to hold or to hunt, and developed their brains—all in incremental steps that helped to preserve the individual or its group. The variations themselves were either inborn or acquired, with beneficial ones propagated through natural selection. Darwin envisioned the winnowing process at work among individuals, nationalities, races, and civilizations, with plucky Englishmen (and their American scions) advancing to the fore.

RESPONDING TO DARWINISM

Darwin's evolutionary naturalism undermined belief in an indwelling spiritual soul, which for many religious people defined the very essence of humanness. After all, it is the mind and emotions that set humans apart from animals. The doctrine that God specially created people by giving them eternal souls carried with it certain implications about the meaning of human life; the theory that humans evolved naturally from soulless animals carried others. From the time of Darwin's *On the Origin of Species*, many individuals found it difficult to switch.

Most, including some evolutionists within Darwin's inner

circle, continued to reject the naturalistic theory of human evolution long after the appearance of *The Descent of Man*. Alfred Russel Wallace, a co-discoverer of selection theory who remained a staunch Darwinist on other matters, became persuaded that an "Overruling Intelligence" created the first humans by ennobling anthropoidal apes with enlightened minds. "Natural selection could only have endowed the savage with a brain a little superior to that of an ape, whereas he actually possesses one but very little inferior to that of the average members of our learned societies," he wrote in a *Quarterly Review* article in 1869 and maintained ever after. Darwin's mentor and friend, geologist Charles Lyell, promptly endorsed this position, much to Darwin's dismay. Although Wallace was not a Christian, he believed in the supernatural and espoused an early version of new-age spirituality.

Late in his life, in the *Fortnightly Review* article "Evolution and Character" from 1908, Wallace could still claim, albeit with some hyperbole, that "all of the greatest writers and thinkers" agreed "that the higher mental and spiritual nature of man is not the mere animal nature advanced through survival of the fittest." Novelist Leo Tolstoy, a mystic Christian who had been excommunicated from the Orthodox Church, proclaimed this viewpoint in Russia, as did the prominent liberal Protestant minister Henry Ward Beecher in the United States. Both embraced evolutionism to a point but maintained that only God could make a soul. Roman Catholic Church doctrine fitfully gravitated toward accepting a similar position. During the late 1800s, British prime minister William Gladstone, an evangelical Protestant, made a point of endorsing the divine creation of humankind. As a general rule,

Orthodox Christian leaders resisted Darwinism on moral and religious grounds, but their fiercest criticism seemed to focus on materialistic theories of human evolution.

Just as some people instinctively rejected the idea of human evolution, others embraced it for reasons that had little to do with science. Materialists, atheists, and radical secularists had long displayed a certain fondness for evolutionary theories of origins, such as Lamarckism (which posits that traits acquired by an organism during its lifetime can then be passed on to its offspring)—anything to dispense with God. Even though Darwin held strictly conventional political and economic views, his theory attracted the usual crowd.

Huxley initially embraced Darwinism in part because it supported his anticlerical agendas for science and society. In America, nineteenth-century feminist leader Elizabeth Cady Stanton welcomed Darwinism as a means to undermine what she saw as biblically based arguments for the subordination of women. "The real difficulty in woman's case is that the whole foundation of the Christian religion rests on her temptation and man's fall," she wrote in *The Woman's Bible*. "If, however, we accept the Darwinian theory, that the race has been a gradual growth from the lower to a higher form of life, and the story of the fall is a myth, we can exonerate the snake, emancipate the woman, and reconstruct a more rational religion for the nineteenth century."

From the conservative end of the political spectrum, the arch-secular social philosopher Herbert Spencer, already an evolutionist before reading Darwin, freely worked Darwinian materialism into his progressive philosophy of social development. As social theorists, Spencer and Darwin became inexorably linked in the public mind during the late nineteenth century. Spencer's many followers typically embraced Darwinism as well. In his *Autobiography*, industrialist Andrew Carnegie recalled the day in the 1870s that his reading of Darwin's *The Descent of Man* and various books by Spencer transformed his life. "I remember that light came as in a flood and all was clear. Not only had I got rid of theology and the supernatural, but I had found the truth of evolution," he wrote. "Man was not created with an instinct for his own degradation, but from the lower he had risen to the higher forms."

For people like Carnegie, Darwinism became a religion, or an alternative to religion. A popular 1883 poster, attributed to London secularist George Holyoake, purported to illustrate the fragmentation of the established British "National Church" into various factions ranging from High Church and Roman Catholicism to dissent and rationalism. In the upper left corner, under the banner of "Darwinism," an ape leads Spencer, Huxley, and other "agnostics" away from the central, umbrella-like dome of London's St. Paul's Cathedral toward a distant cloud of "Protoplasm." A bust of Darwin rises above the cloud. With his great white beard, Darwin could as readily appear godlike as apish. To this day, people still cannot agree on which of the two likenesses is most suitable. In Europe and America, the most prominent agnostics and atheists, from Ayn Rand to Richard Dawkins, are also staunch evolutionists who work Darwinism into the very fiber of their philosophy. The situation is more complex for Eastern religions, which are more readily compatible with an evolutionary view of human development and accept a cyclical view of life.

DIGGING UP THE EVOLUTIONARY RECORD

By the end of the nineteenth century, to the extent that some scientists and many nonscientists continued to reject the theory of human evolution, the lack of known protohuman fossils became a major issue. Other species left a trail in the fossil record, critics charged; why not humans?

Despite cartoons and parlor jokes to the contrary, Darwin and most other late-nineteenth-century evolutionists never claimed that humans evolved from modern apes. Rather, they asserted that all living primates, including humans and apes, had a common ancestor. This branching evolutionary tree should leave a record, but none was known to Darwin or his contemporaries. At the time, evolutionists had only two known types of hominid fossils to work with, both of European origin: the Engis skulls from Belgium, discovered in 1833, and Neanderthal bones from Germany, first uncovered in 1856, with more found later—plus a rich array of Stone Age tools unearthed around Abbeville, France, beginning in the 1830s and whose antiquity was confirmed in 1859. All of these relics apparently dated from an earlier geologic epoch, but neither hominid type exhibited a brain size or structure different enough from those of living humans to constitute a separate species and might have been little more than an earlier race of man. At most, these discoveries pushed the origin of humanity into the distant mist of earlier epochs; they did not link humans to apes. Somewhat different hominid fossils (together with prehistoric cave paintings) soon turned up in France, but these so-called Cro-Magnon creatures apparently were even more like modern humans than the Neanderthals were like modern humans.

Charles Lyell's 1863 book *Geological Evidences of the Antiquity of Man* captured the state of the science in Darwin's day and revealed just how reluctant even Darwin's closest colleagues were to jettison a divine origin for man. In this book, unlike his earlier classic, *Principles of Geology*, Lyell drew on new paleontologic and archaeologic evidence to push back the first appearance of humans into previous geologic epochs, and painted a picture of humanity's gradual cultural development from the time of the Engis fossils, found amid flint implements and the remains of extinct mammals, through the Neanderthal tribes, to the various living races. The first major book of its kind written in an accessible style, *Antiquity of Man* (which went through three printings in its first year) awakened public interest in human geologic prehistory. Yet it held back from endorsing a Darwinian vision of human evolution. The human body may have evolved incrementally from lost types of anthromorphic apes, Lyell conceded, but the human intellect appeared to be the product of great leaps forward. "To say that such leaps constitute no interruption to the ordinary course of nature is more than we are warranted in affirming," he concluded. Darwin felt betrayed. Religious and cultural barriers kept people from accepting the theory of human evolution with more evidence than Darwin offered.

The initial breakthrough in protohuman paleontology came in an unlikely place, due to the near superhuman efforts of a driven Dutch evolutionist, Eugène Dubois. Born in 1858 into a middle-class family in the Catholic provinces of the southern Netherlands, Dubois rebelled against the staid religiosity of his home region and cast his lot with science as the source of truth and progress. As a student, he eagerly read Darwin, Lyell, and Huxley but was particularly inspired by Haeckel's 1873 *The History of Creation.* Haeckel's book opens with a religious challenge: "As a consequence of the Theory of Descent or Transmutation, we are now in a position to establish scientifically the groundwork of a *non-miraculous history of the development of the human race.* . . . It follows from this theory that the human race, in the first phase, must be traced back to ape-like mammals." This became Dubois's mission.

In 1881, when Dubois began his professional career as a assistant in anatomy at the University of Amsterdam, the theory of evolution was widely accepted within the European scientific community, even if opposition remained to the idea that humans evolved. So far as biologists were then concerned, the best proof for evolution came from highly technical studies of morphological relationships between species. The five-finger bone structure of the hand, flipper, and wing of all mammals, revealed either in living or fossil types, was the best known and most persuasive evidence of this type. Given his mission to prove human evolution, Dubois began climbing the academic ladder through a comparative study of the larynx. Speech distinguished humans from other animals, Haeckel stressed in *The History of Creation*, so the development of the larynx should hold a key to human evolution. Although Dubois excelled in this work, he was too restless and ambitious to be satisfied by it.

From his days as a freethinking youth in a traditional Catholic village, Dubois had fantasized about proving human evolution to everyone by finding a fossil link connecting humans to other primates, but he did not know where to look for it. In *The Descent of Man*, Darwin suggested that humans evolved from African hominids. Lyell was less certain. In *Antiquity of Man*, he noted that "anthropomorphous apes" live on the East Indian islands of Borneo and Sumatra as well as in tropical Africa, and urged naturalists to explore both places for the missing link in human evolution. Given his extreme racial views, Ernst Haeckel had little difficulty choosing Asia over Africa as the cradle of humanity. In *The History of Creation*, he hypothesized that a transitional form, intermediate between apes and humans, evolved on a lost continent located off the coast of South Asia. Building on early Aryan mythology, Haeckel proposed that this creature's evolved descendants—first upright walking apelike hominids (which he called *Pithecanthropus*) and later genuine, talking humans—spread across Asia and into Europe, where one branch developed into the Germanic race, which included Anglo-Saxons and the Dutch as well as modern Germans. Haeckel believed that other human races sprang from less developed branches off the same trunk or, perhaps, evolved separately from apes. Dubois followed Haeckel's thinking about human evolution, and it led him to *Pithecanthropus*.

Fortuitously for Dubois, the Netherlands then ruled the East Indies as part of its colonial empire, and it was there that he decided to start his quest. At the time, such matters were more a popular than a scientific concern—something that Jules Verne might write about in a novel. No scientist had ever looked specifically for protohuman fossils. Funding was not available. Professional colleagues thought it ludicrous. Stubborn to a fault, Dubois quit his university post in 1887, signed on as a physician in the Dutch colonial army, and took his young family in search of *Pithecanthropus*, first on Sumatra and later on Java. Most remarkable of all, he found it after nearly four arduous years of looking. Finagling time and support for his project once in the Indies and battling malaria along the way, Dubois examined caves, highlands, and river banks for hominid fossils. In 1891, his workers unearthed a proto-human molar and skullcap from an ancient streambed near the tiny Javanese village of Trinil—the first of such fossils ever found. A humanlike thighbone turned up a year later about fifty feet upstream. Pieced together, Dubois named his discovery *Pithecanthropus erectus*, or "Upright Ape-Man," though most people simply called it "Java Man." The size of its braincase was intermediate between those of humans and apes, but the thigh was clearly made for walking upright, like a modern human. This combination of traits fit Darwin's prediction that walking upright came first in the evolution of humans, with brain growth coming later. The location fit Haeckel's vision of Aryan origins. Dubois's motive shows how science can offer an alternative to religion.

Dubois's discovery created an international sensation. For decades, scientists debated Java Man's place in the evolutionary tree. Some saw it as simply a new species of ape; other considered it fully human. A few openly doubted that the skullcap and thighbone came from the same specimen. Dubois doggedly maintained that *Pithecanthropus* represented an intermediate stage between humans and apes, in part by invoking his concept (derived from a curious mix of Lamarckism and mutation theory) that brain size doubled in each succeeding evolutionary step. By fudging the numbers, he calculated that *Pithecanthropus* had twice the brain capacity of apes and half that of humans—a perfect fit for the sole connecting link between the two types.

Returning to the Netherlands in 1895, Dubois regained a university post but never relinquished control over his fossils. He withdrew them from examination by critics and grew increasingly paranoid. Yet his basic claim that *Pithecanthropus* was a direct ancestor of modern humans slowly gained ground among scholars. Beginning in 1929 with the discovery of similar fossils in China (popularly known as "Peking Man"), *Pithecanthropus*-like fossils turned up with increasing regularity across eastern Asia. Dozens more of the original type were found in Java. Scientists could no longer deny that the protohuman skull went with the more fully human thigh. With further study of more specimens, paleontologists came to see Java Man and Peking Man as older and younger varieties of a single species, much closer to modern humans than to apes, and rechristened them as *Homo erectus.* Researchers would have to look farther back into the fossil record for the missing link between humans and apes. Aryan mythology notwithstanding, the trail led back to Africa.

Late in the summer of 1924, a South African university student brought a fossilized skull to her anatomy professor, Raymond Dart. He identified the skull as coming from an ancient baboon and promptly sought more specimens from the source of the find, a limestone quarry at Taung. Two crates of fossils arrived later that fall. "As soon as I removed the lid a thrill of excitement shot through me. On the very top of the rock heap was what was undoubtedly an endocranial cast or mold of the interior of a skull," Dart later recalled in *Adven-*

tures with the Missing Link. "I knew at a glance that what lay in my hand was no ordinary anthropoidal brain." Hardly larger than that of a modern gorilla, this brain's shape was distinctly more human than that of any living anthropoid, and its orientation suggested that the creature walked upright.

Dart rushed into print with his discovery. "Unlike *Pithecanthropus*, it does not represent an ape-like man, a caricature of a precocious hominid failure, but a creature well advanced beyond modern anthropoids in just those characteristics, facial and cerebral, which are to be anticipated in an extinct link between man and his simian ancestor," Dart announced in the British science journal *Nature*. "At the same time, it is equally evident that a creature with anthropoid brain capacity . . . is no true man. It is logically regarded as a man-like ape." He called it *Australopithecus africanus*, "thus vindicating the Darwinian claim that Africa would prove to be the cradle of mankind." When a chorus of top European scientists dismissed Dart's claims as the rash assertions of a colonial naturalist, the respected Scottish–South African physician-anthropologist Robert Broom, a fervent believer in the spiritually guided evolution of humans, took the lead in defending Dart's position. "In *Australopithecus*," he wrote in *Nature*, "we have a connecting link between the higher apes and one of the lowest human types. . . . While nearer to the anthropoid ape than man, it seems to be the forerunner of . . . the earliest human variety." The debate raged on for years.

Dart made an observation in his initial article that clearly set established anthropologists against him. In focusing their search for early hominids on the tropics, he wrote, they had been looking in the wrong place all along. Luxuriant forests offer a comfortable home for anthropoids, Dart explained, but "for the production of humans a different apprenticeship was needed to sharpen the wits and quicken the higher manifestations of intellect—a more open veldt country where competition was keener between swiftness and stealth, and where adroitness of thinking and movement played a preponderating role in the preservation of species." The savannahs of Africa, and not its jungles, nurtured humanity, and competition was key.

Although Dart never wavered in his belief that *Australopithecus* represented a link in human evolution, Broom became its champion. A respected expert in African fossils, he lobbied for *Australopithecus*'s place in hominid paleontology and actively searched for more specimens. In 1936, he found some in a cave at Sterkfontein, South Africa, and later found more. The new specimens established that *Australopithecus* walked erect and fit into the hominid line, either as an ancestor to modern humans or a branch that became extinct. Broom's fossils comprised two types. Some looked like Dart's original ones, *A. africanus*, while others displayed more robust development and were named *A. robustus.* Two decades later, in 1959, Kenyan paleontologists Louis and Mary Leakey discovered *A. boisei*, an even more robustly developed species of *Australopithecus* in the Great Rift Valley of East Africa, which runs through Ethiopia, Kenya, and Tanzania. "Lucy," a remarkably complete specimen of an apparently older type, *A. afarensis*, was discovered in Ethiopia during the 1970s by a team of researchers co-led by American paleontologist Donald Johanson. Subsequent expeditions to the Rift Valley have uncovered fossils assigned to several more species of *Australopithecus*—

some older, some younger, and some contemporaneous with the previously known types. What once looked like a virtually linear path of hominid progress (from *A. africanus* through *Homo erectus* to *Homo sapiens*) became a increasingly complex and seemingly branching pattern of evolutionary development.

During the late twentieth century and into the early twenty-first century, paleontologists continued to find new types of hominid fossils in East and Central Africa. In 1961, the Leakeys identified a new human species, *Homo habilis* (handy man), from fossil fragments discovered by their son Jonathan. The species supposedly predated and gave birth to *Homo erectus.* Another son, Richard, confirmed the existence of this earlier human type by finding further specimens of it during the 1970s. The Leakeys used newly developed radiometric dating techniques to bolster their claim that the relatively big-brained, tool-using *Homo habilis* had evolved in a land still inhabited by later forms of *Australopithecus*, but that these prior inhabitants ultimately became extinct.

Using modern dating technology, paleontologists estimate that the various forms of *Australopithecus* lived from 4 to 1.5 million years ago, while the first human appeared about 2 million years ago, with *Homo sapiens* coming along only in the past 300,000 to 400,000 years. During the 1990s and after, paleontologists found even earlier hominid fossils (assigned to the genuses *Ardipithecus*, *Orrorin*, and *Sahelanthropus*), with the earliest dated as up to 7 million years old. All of these hominids supposedly walked erect, and none are seen as a common ancestor connecting humans with apes. This would have happened even earlier.

The evolutionary tree for hominids is now as complete as the tree for any type of animal, and it fits a branching, Darwinian pattern. Upright posture apparently came first, presumably because it had survival value in an environment with mixed trees and grassland, then came bigger brains and tool use. Each new hominid fossil discovery generates front-page news around the world. Modern humans remain fascinated by their earliest ancestors. As Dubois predicted, hominid fossils now serve as the best-known and most widely accepted proof for human evolution. Religion has had to adjust to this scientific evidence for human origins or deny it altogether. Many Muslims reject it; Christians and Jews split; and most Eastern religions accept it. The scientific case for human evolution remains the major topic in ongoing debates between religion and science.

TAKING UP THE CUDGEL FOR SCIENCE

If the fossil evidence now shows a gradual, Darwinian evolution of humankind's distinctive big brain, religious believers can still assert a divine origin for human morality and behavior. After all, the soul does not leave a fossil imprint. Most scientists no longer think this way, however. Like Darwin, many of them see human mental and moral attributes—including altruistic behavior and belief in God—as the product of natural forces or, as British science writer Richard Dawkins put it in his popular book of the same name, a "blind watchmaker."

Dawkins's books have become best sellers in the United States and Britain, and his vision of human evolution—or more precisely the vision of it that he popularized from the work of such preeminent late-twentieth-century biologists as William Hamilton and E. O. Wilson—is shared by many evolutionary biologists today. Humans, like all living organisms, "are survival machines—robot vehicles blindly programmed to preserve the selfish molecules known as genes," Dawkins wrote in his aptly named book *The Selfish Gene.* The genes themselves cannot plan ahead or respond to their environment, he stressed. They simply reproduce themselves with occasional random mutations that may or may not assist their survival, and people are the astonishing (or astonished) result. Dawkins found this view of life exhilarating. For him, it frees humans from the burden of purposeful design in nature, which he identified as "the most influential of the arguments for the existence of God." Unlike the controlling purposes of a designing God, Dawkins exalted, "Natural selection, the blind, unconscious, automatic process which Darwin described, and which we now know is the explanation for existence and apparently purposeful form of all life, has no purpose in mind."

Adding urgency to the matter for Dawkins is his conviction that the widespread acceptance of an evolutionary view of life would free society from the destructive consequences of religion. In the preface for his best seller *The God Delusion,* Dawkins described the not-so-subtle ad for a popular BBC-TV documentary called *The Root of All Evil?* Over the caption, "Imagine a world without religion," the ad reprints a picture of the Manhattan skyline with the twin towers still standing. In the book, Dawkins added, "Imagine no suicide bombers, no 9/11, no 7/7, no crusades, no witch-hunts," and so on: "a world without religion," he said, and nothing good is missing. As Dawkins saw it, belief in God and religion once served a Darwinian purpose by helping to bind small tribes together in opposition to others. Now that those small, competing tribes have become large nations or ethnic groups with access to weapons of mass destruction, however, the persistence of this evolutionary trait has become so destructive that it could potentially destroy human society altogether. Only by understanding—a new meme, Dawkins might say—can humanity overcome its now-destructive God gene. Humans should be mature enough to develop and maintain their own ethical code informed by scientific knowledge of its consequences, he argued, rather than rely on an ancient moral code inscribed in scriptural texts of human origin composed in a different era for reasons that no longer apply.

Dawkins finally brought the story to where Reverend Adam Sedgwick feared that it would end more than 150 years ago, when he read the prepublication copy of *On the Origin of Species*. Indeed, from the very outset, what readers generally found most exciting (or frightening) about Darwinism was the prospect that humans evolved from beasts by a naturalistic process that goes back in some material cause-and-effect chain to earliest forms of life. Humans, after all, even those steeped in science or theology, care most about themselves and their own kind. Dawkins expressed this view in starkly dogmatic terms but, except for the bit about genes (which were unknown in the nineteenth century), he did not claim to say anything more than Darwin said in *The Descent of Man* or, for that matter, in *On the Origin of Species*.

Today, Darwin's sketchy social theories have matured by

way of E. O. Wilson's sociobiology and modern evolutionary psychology to become foundational for understanding in the social sciences. Through it, human behavior is reduced to the physical, and people become merely matter in motion with evolved self-consciousness. Meanwhile, as discussed in chapter 3, the rapid march of neurobiology joins evolutionary psychology in linking human mentality with the Darwinian material. Although Sedgwick closed his 1859 letter to Darwin with a lighthearted reference to himself as "a son of a monkey," he never gave ground on a divine source for the human soul. For him and most other Victorians, as for many theistic religious believers today, that was no laughing matter. And science now has much more to say about humans than only their evolution; indeed, it has much more to say about such traditionally religious subjects as sex and gender. We will focus on them and their implications for religion in our next chapter, again turning the lead over to our philosopher.

CHAPTER SEVEN

Sex and Gender

THERE are two creation stories at the beginning of the biblical book of Genesis. In the first, God starts the human race by fashioning Adam and Eve together: "So God created man in his own image, in the image of God he created him; male and female he created them." Then there is another version of human creation in the second chapter of Genesis. It says, "then the LORD God formed man of dust from the ground, and breathed into his nostrils the breath of life; and man became a living being." This time it is clearly only Adam, because soon God realizes that this first man is lonely. So this account states, "So the LORD God caused a deep sleep to fall upon the man, and while he slept took one of his ribs and closed up its place with flesh; and the rib which the LORD God had taken from the man he made into a woman and brought her to the man." The first "Woman . . . taken out of Man."

Much about the human condition is packed in or assumed by these two opening chapters of Genesis. First, humans come in two different kinds, male and female, and this is a fundamental distinction or difference. The Bible does not say "he created tall and short" or "he created black and white." It is male and female. Most people would characterize this distinction as connoting that humans have two different sexes, meaning different physical attributes, particularly relating to reproduction. Average physical size and a lot of other things may differ, too, but often in ways that can be explained in terms of differing biological roles in reproduction. Here, we speak of sexual differences, although the term "gender" includes differences that various societies or cultures attribute to the two sexes, some of which may or may not be founded on actual biological differences. Second, the Bible is quite unambiguous about females' and males' relative statuses. Females and males are humans, and they are distinguished from the rest of creation in both being equally made in the image of God. The use of the word "man" is clearly in the old sense of "human," and more modern Bible translations reflect this. The God of the Jews and the Christians is unlike most of the gods of the time, in being sexless or beyond sex. Generally being in God's image is taken to mean having intelligence and being aware of moral dictates. In the Genesis account, Adam and Eve both sinned by eating the forbidden fruit and both were expelled from Eden. In a moral sense, they appear equal, whether the Genesis account is taken literally or metaphorically. Third, it is difficult to escape entirely the impression that some are more equal than others. The second version of the story has

God creating Adam first and then almost as an afterthought creating Eve to keep Adam company.

For better or worse, this passage sets a pattern that holds through the Hebrew and Christian scriptures. Male-female differences are taken to be real and fundamental. Females are considered as fully human as males, with moral responsibilities and, in the Christian New Testament, with the same opportunities of eternal salvation and bliss with the Creator as males. But at the same time, there is a sense throughout the Bible that males are the dominant, important part of the equation; females, less so and too often weak and dangerous. God makes his covenant with Abraham not with his wife Sarah. It is David who is, above all, the great ruler of the Israelites. It is Jesus who is the savior. And it is Paul and Jesus's other male disciples who preach the Gospel message. Conversely, the serpent deceives Eve first and she leads Adam to sin. Sarah grows jealous of her maid, Abraham's mistress, Hagar, and insists that the girl and her child be sent away, even though she originally had suggested the relationship. For silver, Delilah betrays Sampson to the Philistines. It is Bathsheba's great beauty that distorts David's judgment and leads him to actions that are unworthy. And Paul felt he had reason enough to distrust at least some women that he set up strict rules to govern their behavior: "I permit no woman to teach or to have authority over men; she is to keep silent," he wrote to his assistant Timothy.

But, as always, things are never quite this simple. Looking for real sinners in the Bible, starting with Cain's murder of his brother Abel, men give us lots of examples. Perhaps the greatest biblical sinner is Judas Iscariot, who betrayed Jesus. Like Delilah, he did it for money. And balancing the sins and temptations and weaknesses, there are passages and stories that give an empowering and ennobling perspective on women. There really isn't anyone in the Bible tougher than the Jewish widow Judith from the Catholic and Orthodox Old Testament. She goes off to the enemy camp and ingratiates herself with the leader Holofernes. When he is drunk, she cuts off his head, and that is the end of that threat to Israel. Or for true love and devotion, there is arguably nothing more beautiful or moving in either the Jewish or the Christian scripture than the devotion of Ruth the Moabite for her Jewish mother-in-law Naomi: "Entreat me not to leave you or to return from following you; for where you go I will go, and where you lodge I will lodge; your people shall be my people, and your God my God; where you die I will die, and there will I be buried. May the LORD do so to me and more also if even death parts me from you." Similar accounts of female dignity and agency appear in the Christian New Testament. There is the love of Jesus for Mary and Martha, and their love for him. These were Jesus's dear friends, perhaps his dearest. He does not consider them second-class beings. Remember also that it was Mary Magdalene (who may or may not have been the same Mary) to whom Jesus appeared first after his resurrection and charged to tell the others of this most significant act of the Christian religion. After she recognized him and he comforted her, Jesus tells her: "Do not hold me, for I have not yet ascended to the Father; but go to my brethren and say to them, I am ascending to my Father and your Father, to my God and your God." Thus Mary became the first messenger of Jesus's good news—and a messenger to the brethren, mind you, not just the sisterhood. In short, Judaism and Christianity are far more complex on the nature and status of women than one might suspect from a quick simplistic reading.

ISLAM AND EASTERN RELIGIONS

Complexity on cultural issues is the mark of many great religions, including religions founded on scriptural or authoritative texts. These religions may comfort and give guidelines, but at the same time they can force followers to reevaluate their beliefs and suppositions, making them worry that perhaps there are alternative readings and beliefs. With respect to sexuality, the complexity is true of Islam and of Eastern religions. Founding the essentials of its creation story on the Genesis account, with respect to gender, Islam in many respects parallels historic Judaism and Christianity. "Men have a degree above woman," the Qur'an says: they are the spiritual leaders for their family and community, they are the protectors of women, and they earn the living for their families. But women have equal (if different) dignity. Once married, their sphere is the home. "Paradise is at the feet of the mothers," the prophet Muhammad taught, and Islam generally affords great respect to motherhood, with mothers who die in childbirth seen as martyrs. "Glory to Him Who created all the pairs," the Qur'an says of the female and male couple, and in its shared creation story, Adam and Eve are presented as jointly tempted and jointly sinning (rather than Eve leading Adam into temptation), and there is no mention of Eve being created from Adam's rib. Both female and male are created "from one soul" as a spiritual pair that, while different, completed the other. Religious leadership positions over men are limited to males, however, and female head covering remains the norm during worship. In some Muslim cultures, adult women wear head scarves or burkas outside the home. Nevertheless, as in historical Christianity, heaven is open to both sexes. For the men and women who surrender to and remember God, "the believing men and the believing women, the obedient men and the obedient women," the Qur'an affirms, "for them God has prepared forgiveness and a vast reward."

Moving to Eastern religions, while most privilege men over women, this is rarely clear-cut or categorical. Buddhism offers a case in point. First and foremost, Buddhism is a religion, like certain Western religions or philosophies—Stoicism springs to mind in the ancient world and Spinoza-inspired thought in the modern era—that stresses the importance of freeing oneself from earthly desires, wants, and passions. For the Buddha, craving was a form of suffering (*dukkha*), and sexual passion was one of the chief forms of craving. The aim is to suppress or avoid such feelings. "Letting them go, he will cross over the flood like one who, having bailed out the boat, has reached the far shore," the Buddha said in an early text, the *Aṭṭhakavagga*. In other words, by letting go of sexual passions and other earthly cravings, one is on the way to *nibbana*—the closest one has in Buddhism to reaching paradise (which may in fact be some form of non-being, away from the chain of life and suffering).

Second, in this cultural context, the female seems to come second, and there are strong hints (as in Judaism and Christianity) that men's troubles start with women. As for Catholic priests, celibacy is prized for Buddhists generally and deemed mandatory for monks and nuns. Of course, as with Christianity, it is recognized that this lifestyle is not for everyone. If and when one does marry, things become quite patriarchal. Men are supposed to rule, even though at the same time they are expected to cherish and care for their wives. By some accounts, wives are supposed to get up before their husbands and not relax until their husbands fall asleep. The present Dalai Lama sounds at times as if he is quoting from a Victorian novel, or perhaps from one of today's less-than-sensitive Darwinian evolutionists. He has said (at a conference in Hamburg in 2007), "warfare has traditionally been carried out primarily by men, since they seem better physically equipped for aggressive behavior. Women, on the other hand, tend to be more caring and more sensitive to others' discomfort and pain."

Third, however, in Buddhism women do seem to have equal status as beings, and their merits and achievements are appreciated. Apparently even though a woman can never become a Buddha, it is possible for females to achieve enlightenment. Women can also take on clerical roles. The Buddha's foster mother, his aunt (his own mother died when he was but a few days old), was ordained, and the same was true of his own wife after he had renounced his older ways of living. The Dalai Lama appreciates that women in today's societies can assume roles traditionally taken by men, and he does not as such think this a bad thing. "If the majority of world leaders were women," he observed (in the same Hamburg talk as mentioned earlier) in a remark that accepted gendered differences even as it rejected traditional gendered roles, "perhaps there would be less danger of war and more cooperation on the basis of global concern."

Alone among the major world religions, Hinduism features female goddesses along with male gods—any number of them, some quite important—as well as divine beings that combine female and male aspects. Further, in leading schools of Hindu thought, the Absolute or Ultimate Reality—the final, efficient, and material cause of everything—known as Brahman, is genderless. The perception of some rough sense of gender equality among deities does not necessarily lead to the recognition of gender equality among humans, however. In traditional Hindu culture and law, gender roles follow patterns similar to those historically displayed by the other world religions. Before marriage, for example, females were subject to their fathers; after marriage, wives were subject to their husbands; and if widowed, they became subject to their sons. That is unless upon her husband's death she participated in the traditional Hindu act of sati (suttee) and died upon his funeral pyre—a now outlawed practice that endured in some places into the twentieth century. Further, unlike the other world religions, many branches of Hinduism deny that even the holiest of females could directly attain the highest spiritual state after death, but instead would have to be reincarnated first as a male. On the other hand, of course, matriarchal tribal cultures and belief systems exist that privilege women in this life and the next.

SEX AND GENDER

THE SCIENCE OF SEX AND GENDER OF SCIENCE

Most aspects of the distinctions that various religions traditionally made between males and females had more to do with gender than with sex, and the same was true for many such differences found by ancient natural philosophers and more modern scientists. Historically in this context, and here we are letting our historian take over for a bit, the sciences serve to reinforce religion more than to challenge it. Again, we begin with the Greek natural philosopher Aristotle, whose work in the fourth century BCE profoundly influenced Western scientific thought on this subject for two millennia. What the Viennese neurologist Sigmund Freud famously claimed in the early twentieth century, in the essay "On the Universal Tendency to Debasement in the Sphere of Love" from 1912, about sexual differences—that "anatomy is destiny"—could have been said by Aristotle as well. "In all cases, excepting those of the bear and leopard, the female is less spirited than the male," Aristotle asserted in the *History of Animals.* And the same was true of other virtues like intelligence. In the *Generation of Animals*, a principal work of Greek biological thought, Aristotle went so far as to speak of females as "mutilated males," which resonates with Freud's later concept of women having "penis envy." With this view of women foremost in his mind, Aristotle went on to dichotomize human traits hierarchically, placing supposedly male traits on top. He depicted men, or the male "form," as hot, active, dry, powerful, spiritual, and rational. In marked contrast, he defined the female form as cold, passive, wet, weak, material, and emotional. These were biological traits, Aristotle maintained, which no amount of education or environmental conditioning could overcome. This view supported all manner of cultural and religious subordination of women, fully justifying prevailing Greek and Roman practices with respect to the inferior place of women in ancient society.

In the world of ancient Greek and Roman natural philosophy, Aristotle was not alone in his view of women. Although his teacher, Plato, in *The Republic*, at least accepted the possibility that some women could be equal to men in their abilities—and hence argued that they should be educated at least long enough to allow their individual abilities to appear—on balance he too accepted that men were typically more able to govern than women. Following Aristotle, the Greco-Roman physician Galen, the supreme medical authority of the late-ancient and medieval European and Middle Eastern world, maintained that females were underdeveloped males whose sex organs remained inside their bodies. While maintaining that women were inferior to men, Galen made one important advance over Aristotle. Where Aristotle asserted that the passive female contributed little to reproduction other than a fertile womb that nurtured a male's active seed, observing that children often resemble their mothers as much as their fathers, Galen argued that both parents contribute to a child's inborn traits. Still, even for Galen, it was the man's activating principle that stimulated the female's more passive response in human reproduction. Contributing further to the traditional Western medical view of women, the Hippocratic corpus, another main font of Greek medical thought, defined hysteria as exclusively a female disease. Building on this view, many later Western physicians maintained that, because of their reproductive organs, women were by nature more vulnerable to mental and physical illness than men. The menstrual cycle was viewed as making women unstable, leading some early modern medical writers to argue that women (or at least the upper-class women for whom they wrote) should avoid excess emotional stimulation and physical stress so as to maintain their strength for childbearing and other such "womanly" pursuits.

The medieval synthesis of Catholic and Aristotelian thought, accomplished most famously by Thomas Aquinas in the thirteenth century, reinforced and institutionalized the view that women were inferior to men. Women's supposedly less rational and more emotional nature was used to account for Eve's failure to resist the serpent's temptation, for example, and to explain why more females than males allegedly became witches. It also justified excluding women from the learned professions, such as medicine, law, and the priesthood, and barring them from university education and many pursuits open to men, including natural philosophy. During this period, the rare woman ruler was almost always the daughter, wife, or sister of a king. And so far as the scientific professions were concerned, the situation remained much the same into the modern era. Ample evidence exists to show that, despite such notable exceptions as two-time Nobel Prize-winning physicist Marie Curie (1867–1934), women in all countries and in all cultures were systematically excluded or strongly discouraged from formally training or working in the sciences (except perhaps in some menial jobs) until the mid-twentieth century.

Despite its emphasis on empiricism and looking afresh at old ideas, the so-called Scientific Revolution did little to change the established view of male superiority. The leading seventeenth-century champion of the new sciences in England, Francis Bacon, used gender-laden language to call on active male researchers to "penetrate" the inner nature of nature to discover "her" secrets. Science was increasingly presented as a means to exploit and subject the world in a way that some modern historians portray as perpetuating and expanding the sense of the discipline as a masculine pursuit. The rise of Cartesian mechanical philosophy, with its machine metaphors for both the physical universe and living organisms, is also seen as a more masculine view of nature than the holistic, organic viewpoint that preceded it. For example, the mechanical philosophy gave renewed life to the notion of preformation in reproduction, which had long appealed to Christians. Addressing the age-old question of where individual life comes from, this theory holds that children are preformed in their parents. When the time comes for individuals to be born, they simply expand mechanically from tiny preformed miniatures into babies, much like babies later expand into full-size adults. At first, preformationists tended to envision these miniatures existing in mothers, but with the discovery of spermatozoa in 1677 by Dutch microscopist Antonie van Leeuwenhoek, fathers became the preferred source. Drawing on Leeuwenhoek's discovery and his own subsequent research, Italian biologist Marcello Malpighi promoted the view of male preformationism that held sway in biology until the rise of cell theory in the early 1800s. In 1694, the Dutch microscope user Nicolaas Hartsoeker famously included a drawing of a tiny human form tucked inside a sperm in his *Essai de dioptrique*, which became the icon of the movement. The early-eighteenth-century French priest and natural philosopher Nicolas Malebranche, who sought to synthesis Cartesian and Catholic thought, expanded preformationism to its logical extreme by proposing that every succeeding generation is preformed in miniature and carried from the beginning, somewhat like Russian dolls within their ancestors—miniatures within miniatures—all the way back to Adam.

MODERN EVOLUTIONARY THEORY

Charles Darwin wrote extensively about gender, quite naturally, because in the end, under his theory of it, evolution is all about survival and reproduction. This moves the story into the nineteenth century and takes it further beyond sex, meaning just the differences in anatomy and related characteristics, and more deeply into the realm of gender, meaning roles that are played in society, identifications, and so forth. Along with natural selection, Darwin identified the secondary mechanism of sexual selection, something he made a great deal of when he came to talk about humans in *The Descent of Man*. Sexual selection is the combat and choice among species members for mates—selection by male combat, and selection by female choice. Modeling (as he did for selection generally) from the animal world and extrapolating to humans, Darwin argues that battling among males for females can lead to such natural weapons as antlers, and that vying among males through their physical display for females can lead to such elaborate ornamentation as the peacock's tail.

Generally Darwin thought that sexual selection in humans (as with other mammals) would mostly involve males—particularly males having the most mates. But success in reproduction is not just about quantity; it is also about quality. Offspring must survive and reproduce for mating to make a difference in evolutionary terms. This was where Darwin saw females playing a minor role—and where modern-day Darwinians, like the feminist biologist and UCLA professor Patricia Gowaty, see them playing a major role. No one expects males and females to have identical roles and behaviors. Males in theory can impregnate huge numbers of females and hence have many offspring. Conversely, they may end with no offspring at all. Females, since they must carry, give birth to, and often raise their offspring, can have fewer reproductive pairings than males. Conversely, they are pretty much guaranteed offspring if they want them. At the very least, this means males tend to be less discriminating and females more discriminating. Females must focus on choosing the best partners for themselves and their offspring. Darwin made it clear that he thought sexual selection could go both ways, and in *The Descent* he gave examples of females doing the choosing. There are certainly many examples of this in nature among other species. For offspring survival, Gowaty and others have shown that mate quality and mate variability play at least as significant a role as the gross number of mates.

Being a Victorian male, Darwin was inclined to think that men are swayed by beauty, position, and wealth as much if not more than women are: "Although in civilised nations women have free or almost free choice," he wrote in *The Descent* regarding mate selection, "yet their choice is largely influenced by the social position and wealth of the men; and the success of the latter in life largely depends on their intellectual powers and energy, or on the fruits of these same powers in their forefathers." Darwin certainly thought that, on average and as a general rule, the effects of sexual selection made men and women different from each other. He asserted in *The Descent*, "Man is more courageous, pugnacious, and energetic than woman, and has a more inventive genius. His brain is absolutely larger, but whether relatively to the larger size of his body, in comparison with that of woman, has not, I believe been fully ascertained."

One thing that fascinated Darwin, however, was the more or less equality of sex ratios in humans when fewer men would suffice for reproduction. He could not make up his mind on this one, but he did more or less grasp what most scientists today consider the answer. Natural selection cares most for the individual. If there is an excess of one sex over the other, then from a purely selfish biological point of view, any parent is better off having offspring of the minority type. They stand a better chance of having more offspring in the next generation. This would tend toward equalizing the number, Darwin surmised.

Yet this does not explain why there are two different sexes (with associated gender roles) in the first place. Especially when microorganisms are counted, most species are not sexual. Sex is widespread in the so-called higher animals and plants, however, and sometimes if a species or a group of species includes both sexual and asexual types, the nonsexual types tend to die out first. The most obvious answer for sex is that it enables beneficial new variations to spread quickly through the species. If an organism has a mutation leading to a beneficial good adaptation that aids in its reproduction or survival, such as a form of camouflage, then it is in the interests of the group to spread this adaptation around as quickly as possible, and sex lets one do it. The trouble is that, as with sensible sex ratios, the appeal is being made to the interest of the group over that of the individual. Today's biologists are not being hard-hearted when they question this kind of explanation or opt for the so-called selfish-gene approach to evolutionary forces. This is not a matter of individuals or genes having emotions, but rather of the returns on adaptations and behaviors. Appealing to the interests of the group is too open to cheating. If organism A helps the group and organism B just cares about self, then B is going to get its own benefits plus whatever A hands out. A gets nothing in return and so is at a disadvantage biologically speaking to B. Hence no group-promoting selection and no cozy arguments about sex helping the group unless selection is operating on the group level, too.

And yet there is sex and it persists. Truly there are two problems here: Why did sex get going in the first place? Why does it persist? Looking at the second question first, there are some biologists who think that it persists only because nature cannot get rid of it. The late-twentieth-century American biologist George Williams, for example, was convinced that, biologically speaking, human sexuality is maladaptive, but he was equally convinced that there was no way in which evolution can reverse itself to get rid of the division. Looking at the first question—why sex at all—the British biologist William Hamilton in the late twentieth century observed that parasites and disease are an ongoing problem for organisms, and unless the targets can evolve some resistances, they are doomed. He also pointed out that sexuality shakes up the gene combinations every generation, allowing for new and potentially beneficial variations. He thought therefore that organisms with sexuality could be better adapted to keep up with or stay one step ahead of their parasites, which naturally evolve rapidly, even without sexual reproduction, due to their extremely short reproduction period.

There are two things to note about Williams's proposal: first, it works completely at the individual level. A sexual organism is going to have offspring genetically shaken up, and so the reproductive benefits go to the organism and its offspring and not somewhere else. Second, a good hypothesis should lead to predictions. Hamilton suggested that since disease resistance is so crucial, organisms should signal to potential mates that they have the needed virtues. He thought a lot of sexual selection tended this way, particularly when vertebrates show glowing colors: this advertises their general fitness. This idea expectedly led to a great deal of research about sexual selection and disease resistance. It is probably not the whole story, but it sheds light on the ways that modern evolutionists think about a topic such as sexuality. For all of its virtues, however, certain aspects of Williams's work have been used by some sociobiologists to propose a biological basis for traditional gender roles. Readers need look no further than the work of the "father" of modern sociobiology, E. O. Wilson, who was profoundly influenced by Hamilton's thinking.

Wilson pulled many of these threads together in his 1975 survey, *Sociobiology: The New Synthesis.* Because of its emphasis on adaptations aiding reproduction, the book featured biological accounts of gender-based behaviors. Males naturally tend to spread their ample sperm (including through multiple mates), Wilson suggested, while females tend to conserve their scarce eggs (such as by investing heavily in mate selection and child-rearing). Indeed, some evolutionary psychologists accounted for aggressive behavior by young males as a genetic holdover from a time when it carried reproductive benefit. Among chimpanzees, for example, the most sexually aggressive males produce the most offspring. To some, such explanations sounded like justifications for traditional gender roles, and sociobiology seemed to endorse the social status quo or worse—especially given Wilson's insistence that humans disregard nature at their peril.

THE SCIENCE OF GENDER

Now, with some of the pertinent history of science introduced, we turn the laboring oar back to our philosopher to discuss how religion measures up compared to the science—or, if one prefers, how the science measures up compared to the religion. Part of the difficulty with comparing something like religion on the subject of sex (and gender) and something like science on the subject of sex (and gender) is that these are not two completely independent entities or areas of culture. Although by the time he published *The Descent of Man* in 1871, Darwin's religious beliefs had faded into a form of skeptical agnosticism, he was raised and lived in a Christian country, and it is clear that many of his beliefs—his prejudices—reflected this, and nothing more so than his thinking about sexuality and gender roles in society. As people like Patricia

Gowaty complain, Darwin takes his Christian thinking, reads it into his biology, and then happily reads it out again as confirmation of what he believed all along. Some might think that by now, a century and a half after *The Descent of Man*, all of that Victorian Christianity should have been cleared out of the science. Perhaps so, although critics of today's evolutionary biologists have often made the same charges. Taking this as a caution, not an unsurpassable barrier, at least two pertinent questions arise. The first is about the very fact of sex and gender. The second is about variation, particularly sexual orientation.

Does religion demand male and female? It's hard to say, because it's never really a question that comes up—"he created male and female." Does biology demand male and female? Certainly not, because most organisms do not have sex. They are asexual and reproduce by budding or division and the like. Whatever the cause, it does seem that sex is a powerful tool of evolutionary change, however, and it is hard to imagine higher organisms having gathered together all of the genes that they need without sex.

One interesting question in biology is whether the sex has to be quite the way that it is. Among the barnacles, for instance, it is common to have fully fledged females and little warty males, several of which might attach themselves to one female. In *The Descent*, Darwin himself speculated on the possibility of humans going the way of the bees. "If, for instance," he wrote, "to take an extreme case, men were reared under precisely the same conditions as hive-bees, there can hardly be a doubt that our unmarried females would, like the worker-bees, think it a sacred duty to kill their brothers, and mothers would strive to kill their fertile daughters; and no one would think of interfering." But of course humans have not gone the ways of the barnacles or of the bees, and in some very important sense—the most important sense—males and females are both needed and are equal. As Darwin pointed out, if in some sense males are "better" than females, or the converse, before long, selection will start to even things out, because before long, everyone will start to have offspring of just the one sex, and everyone will lose out.

Religion and biology reinforce each other reasonably well here, as they do also in seeing males and females as having different roles and characteristics. That said, Darwin and most evolutionary biologists down to the present have tended to see males and females as naturally suited for traditional roles—strong men and passive women—and this particularly has been challenged today, as have assumptions about differences in male and female intelligence. And it seems indeed that many of the traditional beliefs are not well taken, or at least are in need of revision. It is true that women are on average smaller, but the fact that they may have somewhat smaller brains does not mean that they are any less intelligent. Writing about the United States in 2009, for example, in their book *The Mathematics of Sex* (2009), American psychologists Stephen Ceci and Wendy Williams reported:

> Roughly half the population is female, and by most measures they are faring well academically. Consider that by age 25, over one-third of women have completed college (versus 29% of males); women outperform men in nearly all high school and college courses, including mathematics; women now comprise 48% of all college math majors; and women enter graduate and professional schools in numbers equal to most, but not all

fields (currently women comprise 50% of MDs, 75% of veterinary medicine doctorates, 48% of life science PhDs, and 68% of psychology PhDs).

Facts like these don't mean that biologists throw out their theories. Rather, like all scientists, they revise them in the light of new facts and findings.

Biological thinking about gender roles has changed dramatically since Darwin's day. This has been partly due to outside cultural influences, but theoretical and empirical developments in science have played a part, too. Thinking about selection from the individual rather than the group perspective encourages evolutionary biologists to see the advantages of being female, rather than looking at females from the viewpoint of the good of the group and not necessarily for the direct benefits to themselves as females. The consequence of all of this is that if, as in modern societies, it is possible to raise and educate females with the quality formerly given only to males, then there are effects, with both males and females readjusting their expectations and lifestyles accordingly.

By this process, people redefine what they mean by "natural." In the mid-nineteenth century, for example, Darwin's great defender Thomas Henry Huxley was strongly against women being allowed into medical schools to become doctors; he thought they should stay at the nursing stage. Today, it is not just the statistics that convince people otherwise. Female physicians and surgeons strike most people as just as natural as male ones.

All of this clearly has relevance to religion, where often much is made of whether or not something is "natural." For example, Catholic natural-law theory claims that morality is all a matter of doing what is natural, with birth control being decidedly unnatural. But just as science changes, so religion also has the resources to change. Even the pope has softened the official church stance on birth control, and countless devout Catholic women use it religiously. Some believe that revealed religion, bound by doctrine and dogma, never changes. As the scholarly Catholic cardinal John Henry Newman said in his "Essay on the Development of Christian Doctrine" (1845), "The Christianity of the second, fourth, seventh, twelfth, sixteenth, and intermediate centuries is in its substance the very religion which Christ and His Apostles taught in the first, whatever may be the modifications for good or for evil which lapse of years, or the vicissitudes of human affairs, have impressed upon it." But as Newman himself pointed out, although some things about Christianity can never change—Jesus as savior of humankind, for instance—that still leaves considerable scope for "modifications." As human understanding of what is "natural" changes, influenced by science in particular and culture in general, so religious practices and customs change. As there are now female doctors, so there are now female Anglican (Episcopalian) priests, when only a few decades ago there were none. Overall, science and religion are in considerable, if dynamic, harmony on such matters.

SEXUAL ORIENTATION

Science and religion do change—maybe not on the fundamentals, but at least on the incidentals. And what may once have appeared fundamental may begin looking incidental. Our historian had a learned senior pastor in the conservative

Christian and Missionary Alliance church who observed from the pulpit that as he grew older there were fewer things he believed, but those fewer things he believed more firmly. Gender roles once seemed fundamental to biologists and theologians alike, and now they seem incidental to many such persons. Scientific evidence and scriptural passages on gender differences have been reinterpreted, with the old understandings now seen by some as naive, misguided, or biased. One topic that often arises when discussing sexuality in the light of science and religion is homosexuality.

Until recently, mainline science and religion almost uniformly deemed homosexuality as abnormal, unnatural, or sinful—and if individual scientists or theologians thought otherwise, they generally kept quiet about it. Until 1973, the American Psychiatric Association classified homosexuality as a mental disorder. Now most behavioral and social scientists and mental health professionals in the United States and western Europe view homosexuality as a healthy variant of human sexual orientation. Given this, there is little point in examining the distant past. We will start a century ago.

More than anyone else, the founder of psychoanalysis, Sigmund Freud, already mentioned in the context of his views on women, set the old standard for America and Europe on homosexuality. He argued that it most commonly was a result of dysfunctional families, in the case of males most famously because of dominant mothers and hostile or absent fathers. This supposedly upset the normal growth of the individual into full heterosexuality, and there was a reversion to an earlier homosexual phase where one is able to fantasize about relations with mother and avoid competition with father. The

Oedipus complex, he called it. For Freud, therefore, homosexuality was a form of immaturity, but he was adamant that it was neither a mental illness nor something that could be altered.

Many researchers in the field today think Freud had it backward. If gay men do more often report close relationships with mother and difficulty with father, then such behavior may arise more from the child than from the parents. Some studies suggest that boys who grow up gay often (but not always) are more likely to elicit love from their mother and put a distance between themselves and their father. This viewpoint suggests that homosexual orientation lies far back, probably prenatally, and may be genetic. Even those scientists and social scientists who disagree about details typically agree that homosexuality is in no way unnatural.

Switching from the physiological to the evolutionary, we find researchers all over the place and none fully persuasive. The percentage of adult gay males seems to be less than 5 percent of the overall male population, a figure that holds across societies. The lesbian ratio is rather less, although this could be a function of the reluctance to self-report: for a female it may be easier to have a straight identity even with a lesbian orientation. Either way, these are small percentages but still large enough to make it improbable that homosexual orientation is either just chance or somehow unnatural. Of course, gays and lesbians can and do reproduce, but (for biological reasons) the figures are significantly less than for straights, and so evolutionists assume that some kind of selection is involved. Orientation could be a by-product of other factors, for instance the genes might be linked to genes for avoidance of schizophrenia. Or there could be direct selective pressures, such as aiding close relatives.

For many contemporary psychologists, sexual orientation and gender indemnity are seen as characteristics that people are born with and cannot fundamentally change. As a consequence, increasingly over the past half century in the Western world, people have seen or felt less and less reason or inclination to make a moral issue of such things. And in step, at least in the West, we see a similar reorientation within religion. The three Abrahamic religions—Judaism, Christianity, and Islam—traditionally condemned homosexuality as sinful and used various scriptural passages to support that judgment. The Jewish scripture (Leviticus 18:22–23) admonishes men: "You shall not lie with a male as with a woman; it is an abomination. And you shall not lie with any beast and defile yourself with it, neither shall any woman give herself to a beast to lie with it: it is perversion." In the Christian one (Paul again): "For this reason God gave them up to dishonorable passions. Their women exchanged natural relations for unnatural, and the men likewise gave up natural relations with women and were consumed with passion for one another, men committing shameless acts with men and receiving in their own persons the due penalty for their error" (Romans 1:26–27).

These prohibitions remained almost universally and dogmatically accepted within those religions well into the twentieth century. Yet change is coming here too. Reinterpretations of the Bible are occurring. For instance, some theologians now see the Hebrew prohibitions as directed against ritual pagan practices rather than against loving same-sex relationships. After all, for whatever it may mean, note the Hebrew account of the great King David's closest male friendship: "the soul of Jonathan was knit to the soul of David, and Jonathan loved him as his own soul" (1 Samuel 18:1). In any case, most modern Christians and Jews take little notice of many Old Testament prohibitions, dismissing them as tied to a different time and place. In the New Testament, accepting that God has made people in different ways, to a growing number of mainline Christian theologians, it doesn't seem compatible with Jesus's teachings to deny them the possibilities of happiness given to others. For these and like reasons, various Protestant dominations in Europe, North America, and Australia, including many Lutheran, Anglican, and Presbyterian ones, now welcome LGBT members, bless same-sex unions, and ordain lesbian and gay ministers. Reconstructionist and Reform Judaism also typically accept homosexuality, and Israel legalized same-sex sexual activity in 1988. Yet differences persist, with many Catholic, Orthodox, and conservative Protestant theologians and church leaders, as well as many Islamic and Orthodox Jewish scholars and teachers, continuing to reject homosexual behavior as unnatural, unscriptural, and sinful.

Turning to Eastern religions, one finds a range that in respects is similar to that found in the Abrahamic faiths. Early Buddhist texts seem not to be overly concerned with such matters, although, as noted, everything is in the context of worrying about desires generally and sexual desires specifically. Later writers can be found condemning homosexuality in various degrees. Although, note also that in Buddhism there is also a context of reaching out to people and not condemning but trying to understand and help through life's journey. Today, the Dalai Lama (in a press conference in San Francisco in 1997) expresses both of these factors: "From a Buddhist point of view, men-to-men and women-to-women is generally considered sexual misconduct. From society's point of view, mutually agreeable homosexual relations can be of mutual benefit, enjoyable and harmless." Many Western Buddhists would endorse the second part of this statement but not the first. That trend reflects an important feature of Buddhism. The Dalai Lama is not the pope, who can lay down the law ex cathedra. Hinduism is similar. Both ancient and modern Hindu texts can be found taking various sides on the issue of homosexuality, from accepting it to condemning it, though clearly the latter type are more typical.

The point with which to conclude this chapter and move on to the next one is that, in the science and religion context, gender differences and sexual orientation raise overlapping issues and concerns. Both science and religion came to these topics with traditions in place, preferences set, and already formulated, culturally laden answers to intensely personal questions. New empirical findings emerged, once-established facts were reinterpreted or discounted, technologies and politics evolved, and new cultural norms and social practices emerged. Science and religion responded to these developments in somewhat similar ways. We can only surmise that responses will continue to develop along parallel lines into the future. And now we pass the baton over to our historian for a deeper look into issues of human genetics and eugenics. As the Greek natural philosopher Heraclitus said (as quoted by Plato in the *Cratylus*) over twenty-five hundred years ago, "Everything flows, nothing stands still."

CHAPTER EIGHT

Eugenics, Genetics, and Playing God

AMERICAN public television broadcast a widely watched documentary series on DNA in 2004, with the episode on genetic engineering provocatively titled *Playing God.* Building on techniques first developed in 1973, the documentary stated at the outset that genetic engineering "enabled scientists to manipulate DNA and transfer it from one species to another, allowing the genetic makeup of plants and animals to be altered in the lab. This work sparked a storm of controversy that continues to this day." By 2000, "playing God" had become virtually a byword for genetic engineering and using genetic knowledge in breeding. Book titles, newspaper headlines, and newsmagazine covers from the period repeated the phrase. At the time, nothing seemed more likely to pit science against religion than these new techniques of genetic breeding. They embodied a power some thought better reserved for God, nature, or chance than scientists, governments, or people.

The documentary went on to highlight several issues. The genetic engineering of bacteria and viruses, the first objects of study, could lead to supertoxic forms that endanger humans and other species. The genetic engineering of field crops, a second major research object, could lead to unhealthy foods and a dangerous dependence on seed suppliers. And lurking behind it all was the specter of a new eugenics—the designed breeding of people. "The most dramatic results of this revolution have yet to be seen," the documentary concluded. "As biologists attain an ever deeper understanding of DNA, the future evolution of plants and animals will be within their grasp. Perhaps one day they will even be able to control the destiny of the human species."

The concern was picked up in countless movies, novels, and plays, such as the blockbuster sci-fi thriller *Gattaca* from 1997, in which people are genetically engineered for their roles in life, and freedom is lost. Many religious believers worried that scientists were somehow usurping God's role as Creator. Some warned that humans crossed divinely designed species lines at their peril. Others feared that spiritual notions of free will and human dignity were at risk. Warnings that genetically modified crops could harm human health, the environment, or farmers rang from various pulpits, particularly in Europe and Africa. In short, drawing on fears of hubris in science and visions of Dr. Frankenstein and his monster, "playing God" had different meanings to different believers—most of them bad. Yet other theologians and religious leaders countered that humans have a moral duty to do good, and that good could be done through the science of genetic testing and gene modification. As new as this controversy seems, its roots go back over a century to science's flirtation with eugenics. That history provides a basis for understanding the current controversy.

THE SCIENCE OF EUGENICS

Eugenics was born in Britain. The term itself was coined by Charles Darwin's cousin, the gentleman scholar Francis Galton, who proposed with respect to human evolutionary development that, to paraphrase his words, what nature does blindly, slowly, and ruthlessly through natural selection, we may do providently, quickly, and kindly through scientific choice. Beginning shortly after Darwin published his landmark *On the Origin of Species* in 1859, Galton began advocating programs to encourage reproduction by those Britons that he perceived as extraordinarily abled (whom he roughly defined as people pretty much like himself) and to discourage it by those that he perceived as particularly disabled (whom he suggested might include habitual criminals and people with hereditary forms of mental illness or retardation).

In fact, what Galton proposed was little more than plant or livestock breeding applied to people, which European royal families had practiced for years, but he made it sound scientific. It is amazing what an Oxford degree and a self-assured, upper-class manner can do to make utter claptrap seem true, especially to those who are inclined to believe it anyway. He must have regretted that, allegedly due to a disease contracted during a youthful dalliance while traveling as a college student in the Middle East, he could not contribute to the cause by siring children of his own, but, if anything, that made him all the more a crusader for it.

Darwin had reservations, in part because no plausible method of genetically transmitting such traits was known. Most biologists then believed that plants, animals, and people could transmit characteristics acquired during their lives to their offspring, so that improving the environment provided a more promising means to improve species than eugenics might. To many in Europe and the United States, this also seemed to be a more Christian approach. Anyone can be redeemed—even a wretch like me! Further, offspring appeared to inherit a blend of their parents' traits, good or bad, not a pure dose of either.

With the rediscovery and rapid acceptance of Mendelian genetics after 1900, however, Lamarckian notions of acquired characteristics lost favor in biology. Now science suggested that parental and other ancestral traits could reappear in children and more-remote descendants without either environmental modification or blending, just as surely as Gregor Mendel had found tall and short pea plants reappearing in fixed ratios. Thus, if there are superior and inferior hereditary traits, and if their impact on succeeding generations is unalterable by environmental influences or blending (both tempting but debatable assumptions under Mendelism), then the scientific case for eugenics could appear compelling—especially if the genes carrying those traits were viewed as operating like simple Mendelian factors. If your father is simpleminded, don't bother with school. If the mother's a thief, lock up the child.

"More children from the fit, less from the unfit," became the motto for a new age of eugenicists in the early twentieth century. Of course, the triumph of eugenics built on a history of greater public acceptance both of hereditarianism in general and of a competitive struggle for existence as the driving force for social and economic progress. It only took a slight twist of reasoning to transpose accepting the natural selection of the fit into advocating the intentional elimination of the unfit. Eugenic marriage restrictions, segregating or sexually sterilizing mentally or physically disabled persons, and even infanticide were widely discussed and actively promoted.

In the United States, the cause was picked up most notably by University of Chicago biologist Charles Davenport—for whom the Carnegie Institution for Science created Cold Spring Harbor Laboratory on Long Island to study the process—and the director of research at the prestigious Vineland Training School for the mentally retarded in New Jersey, H. H. Goddard, who introduced the Binet-Simon IQ test to the United States as a way to make scientific distinctions in mental ability. Davenport, Goddard, and other American eugenicists were soon filling the scientific and social-scientific literature with calls for more children from the abled (so-called positive eugenics) and less from the disabled (or negative eugenics). Many of the nation's state land-grant universities and medical associations jumped on the bandwagon. Robber baron philanthropists and their well-heeled foundations began funding the program in earnest. This was the science of the future: applied genetics at its most promising. The main question was whether the practice would remain voluntary, as Galton largely suggested, or whether it would become compulsory, as some of his disciples urged.

RELIGIOUS RESPONSES TO EUGENICS

Calls for eugenic restraint evoked more titters than protests, so long as those calls remained little more than breeding advice for parents, but the reaction changed when eugenicists began calling for more than voluntary restraint. Then the topic became deadly serious. States began telling people who should marry and who should reproduce, not only in the United States but throughout the developed world. Contraceptives were legalized for some; sexual sterilization ordered for others. Every American state began rounding up the mentally retarded youth—or at least the poor, threatening ones—and locking them in single-sex workhouses during their reproductive years. Many Canadian provinces and Nordic nations authorized and practiced both eugenic segregation and forced sterilization. The new Soviet Union embraced eugenics almost as its birthright. Although not widely practiced there, Cuba, Czechoslovakia, Japan, Hungary, Latvia, Turkey, Yugoslavia, and one Swiss canton also authorized sterilization during the interwar years. Germany, a latecomer to the process, pushed compulsory negative eugenics far beyond any logical extreme after the Nazis gained power in 1933, with the mass extermination first of those deemed disabled and then of Jews, gypsies, and other hated groups. These measures—particularly contraceptives, forced sexual sterilization, and euthanasia—sparked religious reactions in Europe and North America.

Based on its spiritual commitment to the sanctity of all human life regardless of biological fitness, the Roman Catholic Church emerged as the first major organization to challenge eugenics doctrines. Catholics' resistance stiffened as the eugenics movement began advocating sterilization, which the church denounced as violating the "natural law" linking sexual activity with procreation. The pope formally condemned eugenics in a sweeping 1930 encyclical on marriage. Biting essays from the pen of popular British author G. K. Chesterton, a convert to Catholicism, did much to undermine support for eugenics among the educated classes throughout the English-speaking world. Church publications and pronouncements picked up on many of Chesterton's themes and carried them to a wider audience. Wherever eugenics legislation surfaced, local Catholic clergy, lay leaders, and physicians took the lead in opposing it. This opposition proved decisive in defeating eugenics sterilization laws in such heavily Catholic states as Pennsylvania, New York, and Louisiana; helped to stall eugenics legislation in Britain and pre-Nazi Germany; and prevented the spread of eugenics restrictions to the Catholic regions of Europe and the Americas. In Orthodox Christian regions of the Balkans and Eastern Europe, some Eastern Church theologians and leaders raised similar objections to those expressed by Roman Catholics but, especially after the rise of the Soviet Union, typically had less influence over state policies except perhaps in Greece.

Some Protestants also voiced religious objections to eugenics legislation, especially in the American South where fundamentalism was strong and by populist-minded preachers. Evangelist Billy Sunday regularly scoffed at eugenics along with other ideas of "atheistic" scientists, for example, and the evangelical politician and orator William Jennings Bryan began denouncing eugenics in 1923 as part of his populist crusade against teaching Darwinism in public schools. Protestant opponents of eugenics generally did not articulate their position as clearly as Catholics, but it still had an impact. The legislative record is littered with comments by individual Protestants denouncing eugenics as unchristian. Taken as a whole, these objections reflected a view that God controls human reproduction, and neither science nor the state should interfere.

In response, eugenicists actively courted the favor of liberal clerics. In the United States during the 1920s, for example, the American Eugenics Society (AES) formed a committee for cooperation with the clergy that, among other activities, sponsored sermon contests for eugenics. These competitions apparently tapped a reservoir of support because they attracted hundreds of sermons from across the country, most by Protestant ministers. The sermons typically proclaimed biblical authority for selective breeding, linked the spiritual advancement of humans to their hereditary endowment, and invoked Jesus's command to heal the sick and care for the broken. Some traced the eugenic attributes of Christ's pedigree while others foresaw the millennium arriving through the genetic improvement of humanity. God wouldn't have given humans the power to enhance nature if it wasn't for us to use, many maintained. Such reasoning saw science as a gift from God and means for human progress. Harry Emerson Fosdick of New York's prestigious Riverside Church, University of Chicago theologian Charles W. Gilkey, Federal Council of Churches president F. J. McConnell, and other prominent American religious leaders served on AES committees and endorsed eugenic solutions to social problems. At the state level, progressive ministers assumed a visible position lobbying for eugenics sterilization legislation and marriage restrictions.

The pattern persisted in northern Europe. In England, for example, the Convocation of the Church of England and several key Methodist church leaders endorsed the Mental Deficiency Bill of 1912, a highly controversial eugenics proposal. The noted Anglican cleric and writer William Inge assumed a prominent role in the national eugenics movement and limited weddings at St. Paul's Cathedral, where he was dean, to those certified as eugenically fit. Theologically liberal ministers provided similar support for eugenics in Scandinavia, Germany, and other Protestant regions of Europe. This alliance of scientific and religious authority in support of eugenics may have simply reflected the recognition that ministers were effective advocates, but clearly many of them willingly allowed science to inform their spiritual beliefs.

Despite this visible Protestant support for eugenics doctrines in some places, Galton, Davenport, and other leading eugenicists bitterly denounced Christians for opposing scientific progress. In part, these attacks reflected the timing of the eugenics movement, which occurred at the height of the antievolution crusade in America and at a time when secular scholars generally perceived Christianity as historically hostile to science. Eugenicists felt this hostility primarily in the form of vigorous Roman Catholic opposition to sterilization laws, marriage restrictions, and other eugenics measures designed to improve the race. In return, they targeted the Catholic Church for scorn—frequently recalling its persecution of Galileo and casting themselves as modern martyrs for science. More than any issue at the time in Europe and America, eugenics rekindled perceptions of conflict between science and religion.

Yet aside from Japan and Turkey, sterilization laws were only enacted in historically Christian countries. And those two exceptions occurred in secularized states over religious opposition. In Japan, for example, a leading Shinto critic charged that as a divinely created people, the Japanese should not be bred like livestock. Muslims raised similar objections in Turkey.

Judaism offered a different response. Before World War II, many Jewish geneticists, social scientists, and physicians supported eugenics, including those in Weimar Germany before the Nazis took power. Further, in the wake of anti-Semitic pogroms in czarist Russia and eastern Europe during the late-nineteenth and early-twentieth centuries, some Jewish scholars and rabbis saw eugenics measures as a means to rebuild the Jewish people in western Europe, the Americas, and Palestine, where many Russian and eastern European Jews had relocated. Hebrew University historian of Zionism and biology Raphael Falk observed in *The Oxford Handbook of the History of Eugenics*, "The need to secure the continuity of Judaism often gave community considerations priority over the interests and needs of individuals, and rabbis and physicians explicitly linked this tradition to the new 'science' of eugenics." For example, in the 1915 paper "Jewish Eugenics," Rabbi Max Reichler declared, "Jews, ancient and modern . . . have always understood the science of eugenics, and have governed themselves in accordance with it; hence the preservation of the Jewish race." This included eugenics restrictions on who could marry within the community or immigrate to Palestine.

EUGENICS MEASURES

The first task faced by early-twentieth-century eugenicists in every country was to identify those people so disabled that they should not reproduce. Goddard, for example, focused his attention on the mentally retarded and offered a mental age of twelve years as an appropriate minimum level for reproduction. Many early eugenicists targeted repeat criminals, prostitutes, and others who regularly manifested undesirable social behaviors thought to be inherited, though these calls lessened by 1920 as the science advanced. Such physical conditions as hereditary blindness, deafness, and assorted gross deformities were occasionally singled out as grounds for restriction as well. Even an investigatory committee of the American Neurological Association, whose 1936 report sharply criticized eugenic excesses in the United States, nevertheless recommended sexual sterilization for some inherited forms of mental illness and retardation, "disabling degenerative diseases recognized to be hereditary," and epilepsy.

Much of the historical analysis and popular attention given to these matters has focused on compulsory state programs designed to stop breeding by the mentally ill and retarded, epileptics, and criminals. Davenport and Goddard advocated marriage restrictions, forced segregation during the reproductive years, and compulsory sexual sterilization. Other prominent eugenicists offered more drastic remedies, including infanticide. After enough public health experts and physicians echoed these calls, lawmakers and government officials from across the political spectrum responded. Every American state built institutions for forcibly segregating those suffering from various forms of mental or physical disabilities, and thirty-five states enacted compulsory sterilization laws. Hundreds of thousands of people were segregated in the United States and over sixty thousand were sterilized without consent—including over twenty thousand in California alone. Many more were sterilized with only parental consent. Most sterilization operations involved the mentally ill or retarded, but some state programs included criminals, epileptics, and prostitutes. A few Western countries went further, with Nazi Germany using Weimar-era eugenics laws to exterminate Jews and gypsies.

In 1914, Davenport's Eugenics Record Office at Cold Spring Harbor recommended a model program that would sterilize one-tenth of the population every generation. "If the work should be begun during the present decade, it would, in accordance with conservative estimates of future population, require the sterilization of approximately fifteen million (15,000,000) persons during this interval," an ERO bulletin explained. "At the end of the time we would have cut off the inheritance of the present 'submerged tenth,' and would begin the second period of still more eugenically effective decimal elimination." Society surely would support the elimination of this "most worthless one-tenth," the bulletin asserted, and as "public opinion rallies to the support of the measures, a larger percentage could, with equal safety, be cut off each year."

Although no such mass public programs came to pass in the United States, in the now infamous case *Buck v. Bell* (1927), the Supreme Court upheld Virginia's far-reaching compulsory sterilization program for patients institutionalized at the state home for the mentally retarded. Writing for the Court, the famed jurist Oliver Wendell Holmes Jr. explained: "It is better for all the world, if instead of waiting to execute degenerate offspring for crime, or to let them starve from their imbecility, society can prevent those who are manifestly unfit from continuing their kind. The principle that sustains compulsory vaccination is broad enough to cover cutting the Fallopian tubes." Only the Court's sole Roman Catholic member, Pierce Butler, dissented, which Chief Justice (and former U.S. president) William Howard Taft attributed to Butler's religious prejudice. Holmes was a science-minded pragmatist, and Taft was a politically conservative Unitarian, but both believed in eugenics. The Court's sole Jewish justice, the arch progressive Louis Brandeis, joined the majority and, along with Taft, privately praised the decision, which Holmes viewed as one of his finest. Religion and science divided the Court on eugenics much as it divided the nation.

These compulsory programs and proposals represented merely the most notorious product of the eugenics movement. Efforts also were made to educate the public about eugenic breeding—not simply so that people would support compulsory laws for others, but primarily so that they voluntarily would adopt such practices themselves. These efforts included public lectures, popular books and articles, educational movies, traveling exhibits, and classroom instruction. For example, the most widely used high-school biology textbook of the era in America, *A Civic Biology*, featured a section on eugenics that identified mental retardation, alcoholism, sexual immorality, and criminality as inheritable; offered "the remedy" of sexual segregation and sterilization for these disabilities; and gave practical advice on mate selection.

Perhaps the most instructive example of eugenics propaganda from the United States is *Are You Fit to Marry?*, a popular motion picture from the 1910s and 1920s inspired by the case of a Chicago physician who either euthanized or withheld life-sustaining medical treatment from "defective" newborns on eugenic grounds. The movie, which featured the physician in a starring role and was produced by William Randolph Hearst's International Film Service as a commercial venture, explicitly encouraged engaged couples to undergo physical examinations for eugenic fitness, and parents to allow their disabled newborns to die. Yet despite a gloss of scientific objectivism, the movie projected concepts of eugenic fitness and disability that rely heavily on gross appearances. Further, most of those pronounced unfit to marry were disabled more by how people treated them than by any actual physical or mental impairment.

The film's plot centers on a seemingly healthy, wealthy, and intelligent man who carries a so-called hereditary taint. He marries without telling his bride about the taint, and they have a child who displays the defect at birth. The doctor urges her to allow the newborn to die, either by withholding treatment or by lethal medication (the movie suggests both options). She agrees, but only after having a vision of the child's future. In "boyhood," he is taunted for his limp and hunchback by other children. He then asks his mother to read him *The Ugly Duckling*. Unlike in that fairy tale, however, he remains physically disabled in "manhood." The reactions of others also begin to "warp" his mind. After a scene in which his appearance disgusts guests at his mother's mansion, the movie states, "Unable to longer endure the constant humiliation and embarrassment caused by his deformity, he leaves home and becomes a derelict of the street." There he meets further ridicule and rejection, sinks into crime and despair, and sires a brood of similarly disabled children. "God has shown me a vision of what my child's life would be," the mother exclaims at the end. "Save him from such a fate." After dying at the doctor's hand, the baby is shown rising into the arms of an approving Jesus.

NEW VERSUS OLD EUGENICS

This film highlighted issues raised by the history of eugenics, particularly issues of identification, discrimination, and elimination. Eugenicists thought that they could easily identify hereditary disabilities. Once identified, eugenicists discriminated against carriers in an effort to eliminate certain disabilities from the population, following the model of then popular public health campaigns to eliminate certain infectious diseases through compulsory vaccinations. As eugenicist W. L. Champion put it in a 1913 address published in volume three of the *Journal of the Medical Association of Georgia*, such efforts of eugenic identification, discrimination, and elimination promise "to the child of the future the priceless heritage of physical perfection and masterful mind."

In fact, beginning in the 1930s, the scientific case for eugenics fell apart almost as fast as it had come together three decades earlier. First, social scientists reestablished the role of environmental factors in many of the conditions being subjected to eugenics. Then geneticists increasingly realized that, except for such unusual diseases as always-fatal infantile Tay-Sachs disease or the midlife killer Huntington's disease, most genetic traits being targeted by eugenics involve multiple genes and thus could not be effectively propagated or prevented by eugenic selection. Finally, the specter of Nazi abuses discredited compulsory eugenic practices. By the 1950s, eugenics had seemingly been relegated to the dustbin of history or become the refuge of racists and mountebanks. Respectable geneticists no longer publically endorsed it, and its religious critics appeared vindicated. Indeed, many cited eugenics as an example for the ongoing value of religion to regulate science, with science establishing what could be done and religion guiding what should be done. At the time, the atomic bomb offered a better case in point, because eugenics did not in fact show what could be done, but that nuance was overlooked in an era that wanted to restore religious values after the Second World War was fought with new scientific weapons of horrific destructiveness.

Yet even as eugenics retreated, the discovery of DNA's double-helix structure in 1953 and the increasing understanding of how genes impact characteristics and behavior led to a revival of hereditarianism in science and popular culture. Nature indeed trumps nurture, many people again believed. The discovery of genetic engineering in 1973 and its subsequent application to plants and animals revived eugenic dreams and nightmares for humankind. Then in the 1990s the U.S. government launched its ambitious project to map the entire human genome, with the ultimate goal of isolating the genes associated with various hereditary defects, diseases, and traits.

Comparisons are inevitably made between old eugenic efforts for improving human reproduction and new genetic ones. With the Catholic and evangelical Protestant antiabortion movement already in full swing in the United States, religious critics were among the first to draw these parallels, but even some proponents hailed human gene testing and therapy as the dawn of a new eugenics. Unlike the religious critic, however, these proponents typically stressed differences as well: better science underlay the identification process, enhanced legal protection against discrimination, and more sophisticated medical techniques for the elimination of unwanted disabilities. Critics of human gene therapy remained skeptical. To understand the interplay of science and religion on this issue, their claims merit analysis in light of past experiences with eugenics.

THE IDENTIFICATION OF GENETIC DISABILITIES

The advent of human gene testing profoundly changes the process for identifying hereditary disabilities. Of course, the concept of disability still involves the interaction of social and biological factors. Biology may create a physical or mental condition, but only societies or individuals can interpret it as normal or abnormal, ability or disability. However much the modern Western mind favors "objective" scientific definitions of disability, cultural subjectivism inevitably intrudes. This was apparent in eugenics, which claimed the mantle of scientific objectivity but was rife with cultural subjectivism. Lacking better scientific tests, antisocial behavior (such as crime and even poverty) sometimes served as a determinant of eugenic unfitness. In charting the pedigree of disability, Davenport relied on anecdotal family histories. American eugenicists lobbied for eugenics immigration restrictions that barred persons from entry based on their country of origin, with southern and eastern Europeans targeted for exclusion. Goddard's IQ tests reflected deep cultural bias. Populist religious critics, especially Catholics and Fundamentalists with their largely working-class following, highlighted this subjectivity and denounced the entire project as unchristian. Jesus died for all, they would say, not just eugenically fit northern and western Europeans.

Human gene testing promises to diminish subjectivity in the identification of disabilities by isolating the responsible genes and providing tests for the precise molecular detection of individuals carrying them. Indeed, such testing procedures represent the most widespread current application of genetic technology to humans. They are mostly being used in three basic situations: At-risk individuals are tested for certain disabilities that may manifest themselves later in life, such as Huntington's disease and breast or colorectal cancer. At-risk prospective parents are tested for carrying disabling recessive genes, such as testing of Ashkenazi Jews for Tay-Sachs disease and of Mediterranean peoples for beta-thalassemia major, both of which are crippling childhood disorders that frequently lead to early death. At-risk fetuses and embryos are tested for disabilities and other genetic characteristics that run in families or ethnic groups, such as cystic fibrosis. Particularly when in vitro fertilization is used, prenatal genetic screening can extend to a more general examination of prospective traits, with choices of embryos for implantation or fetuses for continuation or termination made on that basis. Gender is one such factor that is easily detected and widely selected for: Do the parents want a girl or a boy?

Yet complications persist in the identification process. First, simply identifying the genetic bases for a particular condition does not resolve its definition as a disability. People differ on what they consider a disease, an abnormality, or a healthy state. For example, some deaf people want a deaf child. Many recoil at the idea of gender selection, and some states attempt to outlaw it, but it is done nevertheless, particularly in China and India. Second, the presence of certain genes, such as BRCA1 and BRCA2 for breast cancer, indicates susceptibility to a disability, not certain affliction. Third, many conditions potentially identifiable through genetic testing vary in the severity of their effect. An individual or fetus with the gene for neurofibromatosis or for tuberous sclerosis, for instance, might die from it, survive with physical disability, or suffer no noticeable impairment. Fourth, just because some disabling conditions (such as colorectal cancer) have an identifiable genetic basis in some does not preclude them from appearing sporadically in others. Finally, there are inevitably errors in testing that produce false positives and false negatives. In these and other circumstances, finding a "disabling" gene does not necessarily indicate the physical manifestation of disability.

Religious critics raise all of these concerns. They represent a broad array of believers who posit an intrinsic value in all human life, from Buddhists and Hindus through both liberal and conservative Christians to Muslims. Catholics, Muslims, and evangelical Protestants have an added objection to prenatal genetic screening because it often leads to the termination of an at-risk pregnancy or the destruction of unwanted embryos. No less a figure than the second director of America's Human Genome Project and subsequent director of the National Institutes of Health, the evangelical Protestant geneticist Francis Collins, raises this concern. As passionately as he supports human gene testing as a means to identify and ameliorate disease in adults and children, he worries about its prenatal use because it could lead to abortions. The former he sees as obeying God's command; the latter he sees as defying it. "When many people think of genetic research, there is a tendency to assume that this means prenatal diagnosis," Collins noted in an interview reprinted in the 1999 book *Genetic Engineering: A Christian Response.* "This is an arena with which I think most Christian physicians are uncomfortable, and I share those feelings." Many Catholic, Muslim, and Evangelical ethicists denounce prenatal screening and actively counsel against its use. From conception, they typically maintain, all human life is sacred and none should be discarded or face discrimination because of inborn, God-given conditions.

THE PROBLEM OF GENETIC DISCRIMINATION

In fact, all forms of human gene testing raise concerns about discrimination. In the case of adults and children, genetic discrimination typically involves persons who are asymptomatic but either have genes that will or may cause a later genetic disability or who carry recessive genes for such conditions. Research indicates that most instances of genetic discrimination relate to insurance and employment, but social ostracizing is also feared by some.

Of course, genetic discrimination is nothing new. The eugenics movement institutionalized such discrimination in various official programs to segregate, sterilize, or stigmatize those deemed to carry hereditary taints. The risk of discrimination posed by human gene testing may be less than under eugenics because, as more testing is done on more people, large numbers of them will be found to have some sort of at-risk genetic factors. Indeed, researchers estimate that all individuals carry five to seven lethal recessive genes. The eugenics movement could single out and discriminate against the "them" who have a family history of hereditary disability, but mostly it looked for such people among the outcasts of their day in prisons, asylums, orphanages, and poor farms. Genetic testing, in contrast, is taking place in the medical offices and hospitals that serve all Americans and is likely to identify persons with asymptomatic but potentially disabling genes among all classes of patients. Thus, there is reason to hope that society will resist widespread genetic discrimination.

Evidence exists to reinforce such hope. In the eugenics

era, governments readily passed laws to discriminate against those who either manifested or carried hereditary disabilities—and no one proposed legislation to protect such individuals from discrimination. Today, there are many such proposals and laws. In the United States, for example, federal law bars both genetic discrimination in employment and health insurance. Similar protections exist in most European countries. Science and religious organizations alike support them.

DEBATING THE ELIMINATION OF DISABILITY

In one important respect, however, there has been little change. Even though antidiscrimination laws reflect new societal sensitivity to people with disabilities, the disabilities themselves remain anathema to many geneticists. For example, the co-discoverer of DNA's double-helix structure and first director of America's Human Genome Project, the arch-secular geneticist James Watson, wrote in his 1992 article "A Personal View of the Project" of finding the genes for Alzheimer's disease, manic depression, and alcoholism so that those disabilities can be prevented. He also dreams of using human gene therapy to enhance intelligence and attractiveness. "The diagnosis of disease-predisposing genes will alter the basic practice of medicine," genetic engineering pioneer Leroy Hood predicted in his contribution to the book *The Code of Codes*. "Preventive medicine should enable most individuals to live a normal, healthy, and intellectually alert life without disease." For these key scientific participants in the effort and many others like them, the Human Genome Project's ultimate goal is nothing less than the eugenicist's dream of eliminating hereditary diseases and enhancing life. They see people taking hold of the evolutionary process and using it for human betterment as a good thing.

The method of doing so remains conceptually similar as well. Eugenicists sought to eliminate hereditary disabilities by using family histories and mental or physical tests to determine procreative fitness. For anyone who did not pass the test, then their remedy was reproductive restraint (including by sexual sterilization). As discussed above, human gene testing provides more precise means to identify persons at risk to transmit hereditary disabilities, but the same remedy persists—restraint. Further, as noted above, the development of both prenatal genetic screening coupled with targeted abortion, and in vitro fertilization coupled with selective implantation provide alternative ways to discourage the birth of persons with disabilities and to encourage disability-free births. Human gene therapy does raise the prospect of correcting disabling genes within the born or unborn, but such techniques remain futuristic and are unlikely to replace selective reproduction as the most practical means of eliminating disabilities.

As the history of eugenics shows, selective reproduction practices can be classified as either compulsory or voluntary. Heightened concern for individual autonomy gives grounds to hope that truly compulsory eugenic practices (such as forced sterilization) can be avoided in the future, at least in Western countries where the worst abuses of eugenics took place. But the history of eugenics also suggests that even so-called voluntary practices can be coercive. The actual event portrayed in *Are You Fit to Marry?* exemplified this: The Chicago physician clearly coerced the mother into voluntarily letting her handicapped newborn die. Local Catholic Church officials successfully demanded his indictment for murder, though a jury ultimately acquitted him after such leading secularists as Clarence Darrow rushed to his defense.

Current practices in reproductive genetics can lead to similar controversies. Prenatal genetic screening is justified as a tool for preventing disabilities and, in the case of fetal testing, this typically means an abortion. Results bear this out. In Sardinia, for instance, where beta-thalassemia major is a relatively common genetic condition, prenatal genetic screening programs have produced striking results. An article in the prestigious *Journal of the American Medical Association* titled "Molecular Diagnosis and Carrier Screening for β Thalassemia" stated that, based on experience with detecting over one thousand fetuses with the condition, one Sardinian genetic testing center reported, "Following fetal diagnosis of homozygous beta thalassemia, almost all couples (99%) so far have decided to terminate the pregnancy." Overall, since the introduction of widespread genetic education, counseling, and screening programs in Sardinia, the incidence of beta-thalassemia major dropped by 95 percent. Notable reductions in incidence due to targeted fetal testing and abortion are reported for other disabling conditions as well, such as for spina bifida in Britain and for Down syndrome in the United States. Some Catholic Church leaders decry this result, but still it happens, even in Catholic Sardinia. Muslim clerics have more clout in Islamic countries, where such practices are more effectively banned, but it undoubtedly happens even there.

Significantly, the genetic counseling that led to such dramatic results in Sardinia was characterized as nondirective. This is standard practice in Western medical ethics, which means deferring to patients' preexisting views. Absent a strong religious objection to abortion, nondirective counseling typically leads parents voluntarily to terminate pregnancies where serious genetic concerns are raised. Experience suggests that this may be true for disabilities with less deleterious effects as well. Some Catholic and Protestant right-to-life advocates, including Probe Ministry's J. Kerby Anderson, have equated these practices to the Holocaust. On the other side, liberal and mainstream Protestants, some Catholics, and most Jews, while concerned about possible abuses, tend to welcome human gene testing and therapy as means to heal the sick and enhance the quality of human life. In his book *Playing God?*, Lutheran theologian Ted Peters takes this position. We should co-create with God to relieve suffering, he argues. That is humanity's moral responsibility. And so long as individual tastes and decisions are involved, it will never be practical or even possible to draw a bright line between relieving suffering or disabilities and enhancing happiness or abilities.

DRAWING LINES FOR SCIENCE WHERE RELIGION MATTERS

Linking the identification of hereditary disabilities with their elimination evoked little vocal dissent in the era of eugenics. Today, however, dissenting voices are emerging, particularly among people with disabilities who liken themselves to members of other minority groups. With different motives and ultimate concerns, both liberal and conservative religious groups in the United States and Europe have rallied to their defense. Both sides typically support legislation to outlaw public and private genetic discrimination and to protect the rights of the disabled, leading to the near-unanimous passage of America's sweeping Genetic Information Nondiscrimination Act of 2008 by a highly partisan Congress. While such laws help keep the government, insurers, and employers from coercing eugenic outcomes, they do nothing to prevent individuals from choosing such results for themselves and their descendants.

In the United States and Europe, where religion is weak or divided on the issue, abortion for any reason remains a key component of a woman's right to choose and in vitro fertilizing and embryo selection have become standard parts of family planning. The increasing use of prenatal genetic screening in these regions makes it more likely that some parents will factor genetics into such choices. Fundamentally secular regimes, such as in China, Australia, and the former Soviet Union, allow similar options. Indeed, as the medical means to choose or change genetic characteristics before birth grow, parents of the future will likely claim ever greater power to prevent or correct perceived genetic disabilities in their children and to add or enhance perceived positive traits.

In stark contrast, where religion is strong and united on the issue, different balances are struck. Most Muslim, Catholic, and evangelical Protestant-dominated countries in the Middle East, Asia, Africa, and South America broadly restrict access to abortion and embryo-selection procedures. By doing so, they reveal a deep divide over the acceptance of scientific developments in reproductive technology that ultimately rests on religious differences. With a few notable exceptions, mostly involving countries such as China and Russia, which became secular during the last century, the geographic and political contours of this division are strikingly similar to the ones over the acceptance or rejection of eugenics during the early 1900s. Like eugenics, human gene therapy has become an issue that divides religious believers and pits some of them against science.

GENETICALLY MODIFIED PLANTS

If designing people is the problem, why all the fuss about genetically modified, or GM, plants? University of California researcher Herbert Boyer created the first transgenic organism by adding antibiotic resistance genes to the DNA of *E. coli* bacteria in 1973, leading to the founding of pharmaceutical giant Genentech three years later. Ever since then, and even more so after the multinational agribusiness conglomerate Monsanto moved into the business of genetically modifying crop seeds in 1983, "Stop GMO Foods" has become an environmentalist mantra almost as familiar as "No Nukes" or "Save the Whales." It echoes from leftist pulpits, in New Age ashrams, and at Jesuit colleges. GMO, of course, is the popular initialism for "genetically modified organisms" and applies to transgenic animals as well as plants, but since there are so few GM animals used for food, public controversy has focused on GM plants.

Within a few decades, genetic engineering transformed medicine. Genentech produced the first so-called biologic drug by genetically engineering bacteria to produce the peptide hormone insulin, which had been in short supply for diabetics and others who need it to live. Previously, insulin was extracted from animal sources. Genetic engineering made it safer, less expensive, and more available, resulting in an immediate medical benefit. Similar methods now are used to produce blood clotting factors, human growth hormones, and other biologics. Except perhaps to demonize drug companies for not selling them even cheaper or developing more of them for rare diseases, once theologians understood that genetically engineered biologics are identical to naturally occurring products, they have not criticized the process even though it involved mixing genes from different species. Quite to the contrary, many hail it as gift from God through science. Let's be frank, any religion worthy of followers cares more about alleviating human suffering than the rights of bacteria.

The public controversy over GM crops proved more contentious than the one over GM drugs, but gradually religious aspects of it have lessened. If fact, the disagreement was never so much between science and religion as it was between some scientists and environmentalists, or among farmers, seed producers, and consumers. After all, the topic here is plants, not people; but as GMO skeptics and opponents point out, not just any plants. The controversy is about plants that farmers grow and people eat. These plants could be made dangerous to the environment or human health through genetic manipulation and, by that manipulation, can remain more the property of the company that engineered them than the farmer who plants them. Environmental protection, human health, and fair treatment for farmers were really the issues that drew clerics and theologians into this debate, not religious concerns about the rights of plants.

Crops are engineered for various reasons. The most widely used genetically engineered crops are modified to resist pests or to tolerate herbicides used to suppress weeds. For example, about 90 percent of the corn and soybeans grown in the United States is engineered to withstand Roundup, which farmers then spray onto their fields to kill everything else, and most of the cotton grown in India, China, and the United States is engineered to resist pests, such as the dreaded bollworm. Scientists certify these processes are safe, but environmental groups such as Greenpeace decry their ecological impact. Genetic engineering has also been used to help some fruits and vegetables stay fresh longer; to create vitamin-enriched grains such as golden rice, which is designed to combat vitamin A deficiency in developing countries; and to allow some crops to grow in salt water. Again, scientists have certified these crops as safe for human consumption, but many people still worry about long-term effects of what some critics call "Frankenfoods." The term itself was coined by Boston College professor Paul Lewis but was famously used by Britain's monarch in waiting, Prince Charles. Popular concerns have virtually killed the market for them in Europe, where foods must be labeled for GM content.

Despite persistent internal debates, few religious groups have taken a stand against GM crops. Even religions with strict dietary rules, such as Muslims, Jews, Hindus, and Seventh-day Adventists, have generally accepted genetically engineered foods unless the inserted genes come from unclean sources, and Amish farmers, who resist many types of modern technology, nevertheless grow GM crops. While Catholic Church leaders, including former pope John Paul II, have raised concerns about abuses associated with GM crops, the Pontifical Academy of Sciences has endorsed the underlying process. In England, the influential evangelical Anglican cleric and Cambridge scientist John Polkinghorne chaired a governmental committee concluding that introducing genes from one organism into another genome, such as done to produce genetically engineered crops, does not violate religious or ethical norms. Social abuses persist in the use of GM crops, many religious leaders conclude, but the process itself can serve humanity.

ALL CREATURES GREAT AND SMALL

In the Islamic Qur'an, Hebrew Torah, and Christian Bible, Adam names the animals, each of its own kind. The First Precept of Buddhism prohibits killing animals. Hindus revere cows and believe that souls currently in human forms can be reincarnated in the bodies of animals. Cats were considered sacred in ancient Egypt. Vegetarianism in Jainism and among Hindus is based on the principle of nonviolence to animals. These and other religions accord special status to animals, or at least to some mammals. Surprisingly, this has not led to many express religious doctrines on the genetic modification of animals or the transfer of genes from one animal species to the genome of another. Scripture being ancient, and genetic engineering being modern, the former does not directly address the latter, but some principles apply.

In general, religious leaders have not opposed the genetic

engineering of any animals except humans, and, even there, exceptions emerge. In England, the Committee on the Ethics of Genetic Modification and Food Use, chaired by the evangelical Anglican priest and former Cambridge University physicist John Polkinghorne, effectively gave a green light to research in the field of animal and plant genetic engineering. In the United States, although Francis Collins has expressed personal religious objection to the concept of humans being co-creators with God, as director of the National Institutes of Health he has not promoted limits on research in the genetic engineering of animals beyond the laws already restricting such work with humans—and these only apply to animal genes being introduced into the genome of humans, not human genes being introduced into animal genomes. This is normally done to create animals capable of producing medical products, but in 1985 U.S. Department of Agriculture researchers did it with the hopes of engineering pigs with lean meat: so much for Porky Pig.

And that's not all folks! Livestock eugenics is freely introducing genes to improve food quality and quantity, with one such animal, a farm-raised Atlantic salmon, preliminarily cleared for sale in the United States and Canada. The British science humorist Douglas Adams satirized these developments in *The Restaurant at the End of Universe*, a sequel to his *Hitchhiker's Guide to the Galaxy*, when the series' hero, the muddled earthling Arthur Dent, met a cow that, to overcome the objections of vegetarians, was genetically engineered to want to be eaten. Their encounter began with the cow introducing itself:

> "Good evening," it lowed and sat back heavily on its haunches, "I am the main Dish of the Day. May I interest you in parts of my body? . . . Something off the shoulder perhaps?" suggested the animal. "Braised in a white wine sauce?"
>
> "Er, *your* shoulder?" said Arthur in a horrified whisper. . . .
>
> "That's . . . the most revolting thing I've ever heard. . . .
>
> "I think I'll just have a green salad," he muttered.
>
> "May I urge you to consider my liver?" asked the animal, "it must be very rich and tender by now, I've been force-feeding myself for months."
>
> "A green salad," said Arthur emphatically.
>
> "A green salad?" said the animal, rolling his eyes disapprovingly at Arthur.
>
> "Are you going to tell me," said Arthur, "that I shouldn't have green salad?"
>
> "Well," said the animal, "I know many vegetables that are very clear on that point. Which is why it was eventually decided to cut through the whole tangled problem and breed an animal that actually wanted to be eaten and was capable of saying so clearly and distinctly. And here I am."

When Arthur relented, the cow called it a wise choice. "I'll just nip off and shoot myself," it said, and then added with a wink, "I'll be very humane."

That, of course, is the issue. What is humane? While some conservative Christian ethicists speak broadly on biblical grounds against introducing the genes of one kind of animal into the genome of another, some mainline Christian and many other religious ethicists accept such processes so long as the intention is to do good and there is no coercion of humans. "I think that the DNA that is in your and my bodies right now is sort of an accident of evolution," Ted Peters stated in his interview for the genetics episode in the highly acclaimed PBS series *Faith and Reason*. "So I hesitate to think of it as sacred, holy, special. . . . If we have the power to alter it in such a way as to make human health better, to relieve human suffering, I think we have a moral responsibility to do that."

Under a theology such as Peters's, genetic engineering, even of people, becomes a gift from God through science. Many geneticists embrace this view as well, but as a gift of a purely human scientific enterprise. Such a view is reflected in one version of the widely parodied diagram of evolutionary progress that (typically moving from lower left to upper right across the page) depicts a fish swimming in water, a fish-like amphibian with legs crawling onto the beach, ever more upright animals walking, and finally a man striding forward. This particular version adds the image at far right of the man rapidly climbing up a spiral staircase of DNA. Absent the then unknown double-helix image of DNA and the sense that the man is doing it so freely, the sketch captures the eugenicists' age-old vision for humanity: using science to control and accelerate evolutionary progress.

Peters and countless other theologians draw the line at genetically engineering humans to serve the state, society, or some other person or entity. The Vatican has laid down this principle as well, as do various Eastern religions. Yet on this, at least for now, most scientists agree. It does not mark a conflict between science and religion as much as show common ground. Compulsory eugenics generally remains anathema to religious and scientific leaders alike. Although potential exists for conflicts to develop between science and religion on the frontiers of applied genetics, unlike in the heyday of eugenics, so far the border has remained remarkably quiet.

Even the recent development of techniques to synthesize

genes from organic chemicals, rather than to take them from other organisms, and place them into genomes has not stirred the water. In his book *Life at the Speed of Light*, the scientist leading this effort, genome pioneer Craig Venter, noted that in 2010, "the response of the press to our announcement of the first synthetic cell . . . followed one track of thought: weren't we 'playing God?'" Yet he added that few theologians joined this chorus and the Vatican newspaper actively discounted this hyperskeptical viewpoint. About the future, Venter wrote in his book with almost religious fervor, "I can envisage that, in the coming decades, we will witness many extraordinary developments of tangible value, such as crops that are resistant to drought. . . . I can image designing simple animal forms that provide novel sources of nutrients and pharmaceuticals. . . . There will be new ways to enhance the human body as well, such as boosting intelligence." Many people, religious and secular, welcome the idea of playing God, or at least doing a bit of co-creating. Perhaps humans can make things better. Of course, there is always the fear that they will make things worse, which leads to us into the issue of science, religion, and the environment. Here we will turn this topic over to our philosopher for a concluding chapter that looks more toward the future than back to the past.

CHAPTER NINE

Living on Earth

IN 2014, Pope Francis issued an encyclical, *Laudato Si'*, arguing fervently that the ways in which humans are treating their home are deeply troubling. The earth is not just a lump of rock and soil covered with water, he asserted, but a creation and gift of God. This sets up a relationship with the earth and imposes obligations. Francis said of all humans that we are called upon "to accept the world as a sacrament of communion, as a way of sharing with God and our neighbors on a global scale. It is our humble conviction that the divine and the human meet in the slightest detail in the seamless garment of God's creation, in the last speck of dust of our planet." He continued, "We are not God. The earth was here before us and it has been given to us. This allows us to respond to the charge that Judaeo-Christian thinking, on the basis of the Genesis account which grants man 'dominion' over the earth (cf. *Gen* 1:28), has encouraged the unbridled exploitation of nature by painting him as domineering and destructive by nature. This is not a correct interpretation of the Bible as understood by the Church."

GLOBAL WARMING

Pope Francis has two targets in his sights. The first, most obviously, is the present state of the environment. Not just the ravages of open-cast coal mining or the pollution of the seas by plastics and other human-made materials, but above all global warming, the apparently nigh-inevitable increase in the earth's temperature thanks to human use of fossil fuels. Take this concern back to its start: the sun. It gives off heat, and it is this that makes everything possible on the earth. Some of the sun's heat bounces back off into space. This is particularly the case in the Arctic and Antarctic, where sea ice and ice sheets reflect the rays. Some of the sun's heat gets absorbed by the planet. Think of how the black tarmac heats up in the summer. And some of the heat gets trapped by the earth's atmosphere—the so-called greenhouse effect. This is the crucial part of the tale. The atmosphere is a combination of different gases, notably oxygen and nitrogen, but also traces (or more) of other gases, some of which, like water vapor, absorb heat and act like a blanket over the earth. Some of these "greenhouse gases," methane notably, really trap the heat, but they do not create a long-term problem because they are unstable and break down within a matter of years. However, any increase in atmospheric carbon dioxide (CO_2) can persist for centuries or even longer because of the complex global carbon cycle. So the more carbon dioxide in the atmosphere, the more trapping and the more heat for a very long time. Animals respire—they inhale oxygen and they exhale carbon dioxide—but plants need carbon dioxide, and so overall a balance is achieved. Humans are and always very much were part of the equation. Because of the greenhouse effect—and this shows that it is not in itself necessarily a bad thing—the earth is up to 60 degrees Fahrenheit warmer than it would otherwise be. The temperature allowed humans to evolve from cold-blooded species.

The theory that carbon dioxide generated by the burning of fossil fuels causes global warming developed over the past 150 years. During the mid-1800s, British physicist John Tyndall first showed that CO_2 and other so-called greenhouse gases trap radiant heat from the sun and emit it into the atmosphere. Early in the 1900s, Swedish researchers Svante Arrhenius and Nils Gustaf Ekholm found that atmospheric carbon dioxide has had more impact than any other greenhouse gas on changes in global temperature and that, going forward, increasing its concentration by burning fossil fuels should warm the climate. Ekholm estimated that tripling the amount of CO_2 in the atmosphere could raise average temperatures by up to 9 degrees Celsius. Data collected by British engineer Guy Callendar in the 1930s showed that, since the dawn of the coal-fired Industrial Revolution in the late 1800s, the climate had in fact already warmed about 0.5 degrees Celsius as carbon dioxide levels rose, and that both trends correlated with fossil-fuel use. In classic scientific fashion, the emerging physical evidence confirmed the hypothesis that increased atmospheric carbon dioxide caused higher average global temperatures.

At no time was anyone concerned about these developments. Coming from Sweden, Arrhenius and Ekholm actually welcomed the prospect of global warming. Ekholm even suggested accelerating the process by exposing and burning shallow seams of coal to release more CO_2 into the atmosphere. During the mid-twentieth century, some Soviet scientists took up this cause as a means of geoengineering to thaw the Siberian permafrost for farming and melt the Arctic ice cap for shipping. Climate change, they believed, would help the Soviet Union win the Cold War.

Over the century, however, researchers began expressing alarm over the impact of carbon dioxide emissions on the global climate. In 1961, Mikhail Budyko, a Russian who pioneered bringing the application of quantitative methods to climatology, warned that waste heat from energy generation could render the earth uninhabitable and, eleven years later, he released a model suggesting that the warming attributable to rising levels of atmospheric CO_2 would melt the Arctic ice cap and significantly raise sea levels by 2050. In Hawaii during the 1950s, Scripps Institution of Oceanography chemist Charles David Keeling began systematically measuring atmospheric CO_2, finding a steady rise over time correlated with increased burning of fossil fuels. Scripps director Roger Revelle used this so-called Keeling Curve in a 1965 governmental report to predict that the amount of carbon dioxide in the atmosphere would increase 25 percent by 2000, potentially causing "marked changes in climate." This report, which was the first U.S. governmental statement on global warming, led President Lyndon Johnson to warn Congress in 1965 that "a steady increase in carbon dioxide from the burning of fossil fuels" was altering the composition of the atmosphere on a global scale.

Unfortunately, intelligence is a two-edged sword. Thanks to technology, people are adding carbon dioxide to the atmosphere and removing the natural filters, partly by deforestation and overwhelmingly by the use of fossil fuels (about 90 percent). Around 1750, at the dawn of the Industrial Revolution, as ice-core analysis has suggested, CO_2 levels in the atmosphere stood at 280 parts per million (ppm). For the four hundred thousand years before that, researchers estimate the levels varied between about 200 ppm during ice ages and 280 ppm during warmer interglacial periods. Today, due in large part to industrialization, the level is over 400 ppm. Even more significant is the fact that most of this change is recent. Humans are pouring CO_2 into the atmosphere at six times the rate that was happening in 1950. In 1970 even, the amount of CO_2 in the atmosphere was only 325 ppm. So what does this all mean? Ask first about today. Already the earth is warmer than it was. Since 1900 the increase has been 1.5 degrees Fahrenheit, and 1.0 degrees of that has been since 1970. And already the trickle-down effect is becoming apparent, although "trickle" is hardly the right word. Summer after summer, scores for the record books show each summer to be the hottest since records were first taken. Droughts across the United States, in California for example, almost no longer constitute news. Alterations in the wind currents in the atmosphere do not in themselves necessarily bring on superstorms like Hurricane Sandy on October 29, 2012—which left a hundred people dead and $70 billion in damages—but they intensify them and make their courses more erratic. For obvious reasons, a warmer sea surface means more evaporation means more potential rainfall. The sea off the coast of the Eastern United States, where Sandy was to strike, was 5 degrees higher than normal, and global warming had a significant causal input. Because of the loss of sea ice, temperatures are rising fastest in the Arctic, with the annual average rising over 2 degrees Fahrenheit from 2010 to 2015.

The sea rises seem inexorable. They are now going up at well over an inch a decade and more ominously going up now at double the rate for the twentieth century taken as a whole. The earth's ice caps are melting at a frightening rate. The Greenland ice sheet, the size of Mexico, doubled its melt rate from 2009 to 2015! If it all goes, the seas could rise twenty feet. The West Antarctica ice cap is also showing signs of increasing instability and rapid melting. If it goes, look for another twelve feet of sea rise.

If the present is bad, the future is worse—and frightening. Think lifestyle and foodstuffs. When it is hot, people work less because they are unable to work more. Scientists can even quantify these sorts of things. Above about 80 degrees Fahrenheit society loses about 1 percent of productivity for every degree that the temperature rises. And of course there are all sorts of related illnesses, not to mention serious questions about mental efficiency when CO_2 levels are high. In the case of food, a major worry is the acidification of the oceans, an effect of increased CO_2 levels. Right now, oceans are acidifying at a rate ten times that of fifty-five million years ago, when a major marine extinction event occurred. Coral reefs are suffering dreadfully, and in turn the marine life dependent on them suffers and declines. There are simply fewer fish in the ocean for us to haul out and eat. On land too there are harsh effects. Crops like soybeans and corn simply cannot handle higher temperatures. A 5-degree rise in temperature could mean, at a minimum, over a third of fertile cropland world-wide is lost. In 2011, the Royal Society in London concluded a report by saying it feared "large losses in biodiversity, forests, coastal wetlands, mangroves and salt marshes, and terrestrial carbon stores supported by an acidified and potentially dysfunctional marine ecosystem." And a three- or four-foot sea-level rise by century's end is almost to be expected. Many models now predict even greater changes. If this happens, it will be a disaster for coastal areas.

CHRISTIANITY'S GUILT?

Who is to blame? The pope's second target, less obvious but no less important, picks up where the last chapter ended. It is claimed by some scholars that a root cause of environmental problems, and a major factor in the denial by so many of the threats to the environment, lies in religion. And not just religion but they say one dominant Western religion, Christianity. The historian Lynn White Jr., in a celebrated article in the journal *Science* in 1967, set the scene, arguing that thanks to their religious beliefs, Christians feel justified in thinking that the world exists for their purposes and that they are allowed to do with it what they will. Christians believe they have a Creator God who made everything, White reasoned. In the Old Testament, the prophet Nehemiah is pretty good on these sorts of things: "Thou art the Lord, thou alone; thou hast made heaven, the heaven of heavens, with all of their host, the earth and all that is on it, the seas and all that is in them; and thou preservest all of them; and the host of heaven worships thee." In the New Testament, Paul is a reliable source, as in the Epistle to the Colossians: "For in him all things were created, in heaven and on earth, visible and invisible, whether thrones or dominions or principalities or authorities—all things were created through him and for him." This creation is entirely God-centered—"created through him and for him"—and humans have a special place. Thus Psalm 8:

> When I look at thy heavens, the work of thy fingers,
> the moon and the stars which thou hast established;
> what is man that thou art mindful of him,
> and the son of man that thou dost care for him?
> Yet thou hast made him little less than God,
> and dost crown him with glory and honor.

This sort of thinking, White argued, leads Christians to believe they have the right to do with the world what they will. Back to Genesis and the creation of Adam and Eve: "And God blessed them, and God said to them, 'Be fruitful and multiply, and fill the earth and subdue it; and have dominion over the fish of the sea and over the birds of the air and over every living thing that moves upon the earth.'" Related to all of this is a philosophy of change and action that is rooted deeply in the theological underpinning of Christianity (and the Judaism from which it emerged), White added. People should use their God-given powers of reason and develop science and technology, imposing solutions on the world, and thus solving life's problems and pointing toward ever greater comfortable and threat-free living. In this sense, White argued in a historical analysis that fit early-modern Protestants better than Catholics, Christians put their faith in the ever present possibility of progress. In White's words (in the *Science* article): "Our daily habits of action, for example, are dominated by an implicit faith in perpetual progress which was unknown either to Greco-Roman antiquity or to the Orient. It is rooted in, and is indefensible apart from, Judeo-Christian teleology."

White would not have been at all surprised at the reactions of many Christians today toward the global-warming crisis. The conservative talk-show host Rush Limbaugh is on public record (on his show in 2013) as saying, "See, in my humble opinion, folks, if you believe in God, then intellectually you cannot believe in manmade global warming. . . . You must be either agnostic or atheistic to believe that man controls something that he can't create." In 2005, the conservative evangelical Cornwall Alliance for the Stewardship of Creation, in their "Biblical Perspective of Environmental Stewardship," stated, "Just as good engineers build multiple layers of protection into complex buildings and systems, so also the wise Creator has built multiple self-protecting and self-correcting layers into His world," adding, "Therefore a Biblical theology of Earth stewardship will recognize the superintending hand of God protecting the Earth. Particularly when it is combined with . . . observations about the resiliency of the Earth because of God's wise design, this ought to make Christians inherently skeptical of claims that this or that human action threatens permanent and catastrophic damage to the Earth." So why worry about carbon dioxide emissions or deforestation? God will protect Christians to the end of time, which may come pretty soon.

Pope Francis wants nothing of this kind of thinking and, to be fair, things are not quite as simple as White and followers suggest. There is biblical warrant for alternative readings and interpretations of human relationships toward and responsibilities for the world, God's creation. As one might expect from a nomadic people, much is couched in terms of shepherding and caring for flocks, rather than outright exploitation. Apart from anything else, the Bible suggests in Psalm 24, people only have the earth on loan:

> The earth is the Lord's and the fulness thereof,
> the world and those who dwell therein;
> for he has founded it upon the seas,
> and established it upon the rivers.

And that means that God sets the rules, for as it says in Exodus 23, "For six years you shall sow your land and gather in its yield; but the seventh year you shall let it rest and lie fallow, that the poor of your people may eat; and what they leave the wild beasts may eat. You shall do likewise with your vineyard, and with your olive orchard."

In the same spirit, in *Laudato Si'*, making reference to the Genesis account of the Fall, when the first humans rebelled against God's commands, the pope wrote:

> The creation accounts in the book of Genesis contain, in their own symbolic and narrative language, profound teachings about human existence and its historical reality. They suggest that human life is grounded in three fundamental and closely intertwined relationships: with God, with our neighbor and with the earth itself. According to the Bible, these three vital relationships have been broken, both outwardly and within

> us. This rupture is sin. The harmony between the Creator, humanity and creation as a whole was disrupted by our presuming to take the place of God and refusing to acknowledge our creaturely limitations. This in turn distorted our mandate to "have dominion" over the earth (cf. *Gen* 1:28), to "till it and keep it" (*Gen* 2:15). As a result, the originally harmonious relationship between human beings and nature became conflictual (cf. *Gen* 3:17–19).

Pope Francis too wants to make more of "dominion" than a simplistic reading would suggest. In demythologizing nature, Judeo-Christian thought "emphasizes all the more our human responsibility for nature," Francis explained. God has put people in charge—they have dominion over nature—but that means they have the obligation to care for it and maintain it. It is appropriate to think of the planet as part of the family. "Saint Francis of Assisi reminds us that our common home is like a sister with whom we share our life and a beautiful mother who opens her arms to embrace us," Pope Francis noted of his chosen namesake.

What about progress? Pope Francis is far from being a Luddite, but he does not show quite the unalloyed enthusiasm for progress that might be expected from White's account. He certainly does not think it a universal cure-all. And, actually, history suggests that there are reasons for the pope's caution, and he is neither against progress as such nor overwhelmed by its prospects. White is certainly right in saying that the Greeks and Romans had no true notion of progress. Plato, for instance, believed in a world that was infinitely old and that all change was limited and more or less cyclical. Roman Catholics, at least into the Middle Ages, retained a general view that life on earth was not progressing. Their hope was in the afterlife.

Progress, the idea that humans can improve things through their efforts (whether aided by God or through their unaided talents), was very much a child of the Protestant Reformation in the sixteenth century and the Enlightenment that followed, beginning in the seventeenth century. It was then, thanks to science and technology and more (like foreign travel) that increasing numbers of thinkers believed that they could improve society, perhaps indefinitely. It is undeniable that in some sense the idea of progress comes from Christianity, particularly Protestantism. But it is misleading not to point out how, at the very least, notions of earthly progress challenge Christian philosophies of change. It comes right up against beliefs in Providence, the idea that God alone can save us through his supreme sacrifice. Human efforts and results count for naught. Right up to the present, Christians confront this tension.

It is true that as the years went by and as science and technology grew ever more powerful, many Christians came onside, at least in part, with the notion of progress. And historian Peter Harrison has shown how some deeply religious Reformation-era natural philosophers, such as Francis Bacon and Robert Boyle, envisioned virtually limitless progress through science-restoring, pre-Fall powers to humans with the advent of Protestantism. More generally, particularly with the widespread acceptance of theistic theories of evolution in the nineteenth century by mainline Protestants and some Catholics, there was recognition that the world works according to God-given natural laws, that in some sense it is developmental and has been unfurling in a seemingly progressive manner that has led to human beings. To many, this implied that people have an obligation to run their lives accordingly by striving for further improvement. Progress was and is possible. Yet, there was and is always the background supposition that without God people are nothing. There was and is the supposition that whatever people do should be in accordance with God's will and moral laws. Simply doing things for personal gain is not acceptable. Overall, trying to make for a better world is not just permissible, but a Christian duty.

This obviously accounts for the pope's hesitancy. Progress without God is going to go nowhere. In this respect at least no one should blame Christianity for all science-and-technology-produced problems. But going back to global warming, what is to be done, and can what is to be done be done in a Christian fashion? It should be noted that it is not just a question of maintaining the status quo and not making things worse, because if the status quo continues, things are going to become much worse. That means that industrial countries have got to cut carbon emissions dramatically.

SCIENTIFIC SOLUTIONS?

Today, as well known as the problem of global warming is, so are many of the proposed solutions. Some are utterly unrealistic. One thinks for instance of grandiose geoengineering plans proposed by some scientists to put mirrors or shields in space to deflect the sunlight. No one knows how properly to do something like this, and even if they did, it would cost a fortune. Better by far to stay down here and work with what is known. Among these, first and most obviously, especially to anyone who lived through the oil crisis of the 1970s and the effective responses, one practices conservation. In a cold snap, don't fiddle with the thermostat. Put on a sweater. Build houses that are a lot more heat efficient than they have been hitherto. Use devices (like light bulbs) that use less energy, and so forth. Already huge amounts of shopping can be done through the Internet, avoiding the need for going to stores. Coupled with this are such things as more efficient use of public transportation. It always amazes anyone who is used to European cities to find in America how little people are willing to join together to share transportation. Certainly things can and must be done about internal combustion engines, starting with ways to make them more fuel efficient and going on to ways to make them more and more redundant, or at most adjuncts to other motors, like electric-powered machines. Yet all this is but a small drop in a very large bucket.

Second, dependency on coal must be reduced. It is expensive (if not impossible) to extract the carbon dioxide emissions from coal-powered electrical power plants. And there are obvious alternatives, or at least partial alternatives, to coal for producing electricity, starting with solar power and going on to wind power. One thing on which everyone agrees is that the cost of solar power has declined dramatically over the past decades and continues to do so. Wind power too is getting a lot more efficient and will and can provide much of our electrical needs. In Denmark already wind power is providing over a third of the nation's electricity, and with huge efficiency the cost of power is half of what is needed for the profitable running of coal plants. Nuclear power is more problematic. On the one hand, it does not burn fossil fuels and so it does not contribute to global warming. On the other hand, there remains the short-term environmental risk of radiation releases and the long-term puzzle of what to do with the radioactive wastes. The 2011 earthquake in Japan, which led to devastating releases at the Fukushima nuclear plant, spurred countries like Germany to eschew all nuclear power. However, clearly coming to a decision in other countries depends on the availability of nuclear fuel and the unavailability of other fuels. For some environmentally concerned scientists and engineers, nuclear power remains a better option than coal for generating electricity but far from ideal. Natural gas is another cleaner alternative to coal but raises concerns about the impact of fracking.

Third, briefly, there are other potential sources of power, for instance using underground heat for power generation or directly going to homes to heat water or living spaces. Already, a third of the electrical power in Iceland comes this way. More controversial is the use of plant products to make synthetic fuels equivalent to natural hydrocarbon fuels. On the one hand, one has the problem of pollution, as with natural fuels. On the other hand, to produce "biofuels" one needs cropland and this puts pressure on the availability of land for growing food. One promising line of research is to use not the food products of plants but their nutritionally useless by-products, like leaves and stems. Also, expectedly, work is being done on genetic modification of plants more efficiently to produce material ready to be converted into usable fuel.

None of this work is directly related, for or against, to

Christianity in particular or religion in general. An outright atheist can be as concerned about global warming as can be theists, and as optimistic (or not) about humanity's chances of change. One suspects that many atheists think they are going to be more concerned about global warming because for them this is the only life they have and so they had better get it right. Richard Dawkins, for one, thinks that secular humanity offers a glimmer of hope. This didn't come about by design, he characteristically adds. It was pure luck. Humans developed brains to survive and reproduce more efficiently than their competitors, and one result—one by-product almost—is that they can see into the future, they can plan and act, and they can change things for the better. "Better," that is, in the sense of what they want, not better in some absolute sense, nor even really in some biological sense. This wasn't always the case. It is pretty clear that humankind's Pleistocene ancestors wiped out huge numbers of animals, and early agriculture wasn't much better, either. Slash-and-burn leaves the land devastated. But now humans can plan and change things. "The only hope lies in the unique human capacity to use our big brains with our massive communal database and our forward simulating imaginations," Dawkins concludes in a lecture given at the Royal Institution in London in 2002. This is one view from science, or at least from one widely followed science writer who desperately (some would say sentimentally) wants civilization to survive. He would have no problem in principle with using the methods available to counter global warming.

Nor would someone like the pope, for all that he expresses caution about technology and a blind belief in progress. As a Christian he obviously would have no worries about putting on a sweater during a cold snap at the Vatican, nor would he mind solar panels at his country summer home, Castel Gandolfo. Actually, his predecessor, Pope Benedict, had solar panels installed at the Vatican, and there are plans for a biomass facility at the pope's summer home. The present pope is ahead of many on these issues. He has no problems in principle with genetic engineering, even though some would argue that it is necessarily wrong because in some sense people are playing God. The pope's response would be that in some respects people are expected to play God, given that they are made in His image. It is all a matter of what "some respects" actually covers. One suspects that the pope would expect Christians to take action, by seeing that nature's bounties are distributed evenly, and if sacrifices are to be made, that they too are distributed evenly. More broadly, one has to think not just of humans but of the rest of creation. Notoriously, wind generators can be deadly for birds. Issues like these must be part of the package. If it is an important design problem to make more efficient use of the wind, it is also an important design problem to make machines that are safer for wildlife. One major drawback of nuclear power plants is the huge amount of water they need for cooling. Christians and atheists alike would be uncomfortable with rivers of boiling water emptying into the homes of unsuspecting wildlife.

PAGANISM AND EARTH WORSHIP

So what about other religions and their implications for the environment? In his analysis of the issue, White (in his *Sci-*

ence article) noted that Eastern religions have no Creator God and hence do not make humans central to God's purposes. Indeed, he asserted that Buddhism "conceives of the man-nature relationship as very nearly the mirror image of the Christian view." White also pointed to the theology of Francis of Assisi and what White called "the Franciscan doctrine of the animal soul" as advancing environmentalism in an alternative Christian context. Yet as a Californian who was living in the 1960s, he could not resist mentioning paganism as well.

For many people, paganism conjures up a crowd of people prancing around stark naked in the woods by moonlight. Often, as with the case of the title character of the Verdi opera *Falstaff*, stag's antlers are involved. There are folks of this ilk. Today, they usually like to be referred to as "Neo-Pagans." White, however, thinking more broadly, probably meant to include people who would be very surprised to be considered pagans at all. Negatively, paganism is associated with religion beyond the Abrahamic religions—Judaism, Christianity and Islam. Positively, it is associated with convictions that humans should respect and perhaps even worship nature. In this rather loose and broad sense, such thinking is widespread, whether it be incorporated into a movement like deep ecology or ecofeminism or biodynamic agriculture, or simply the background assumption of many Sierra Club members or Greenpeace donors.

The central idea of this sort of paganism is that of a world soul. The earth is an organism, with life as vital and as abundant as that of any creature that lives on or below its surface. Indeed, it is humankind's collective mother in some very real and literal sense. This "Mother Earth" is (at least it was before humans ruined it) the source of all that is good and life-giving: the streams and lakes, the marshes and fens, and of course the sea; the fields and the crops and the woods as a source of fuel and building material; the minerals of increasing importance, like iron, copper, tin, and obviously gold and silver; the animals and birds and fish, sources of labor and of food. Moreover, like a living being, the earth is seasonal, starting to life in the spring, then flourishing in the summer and giving foodstuffs up to and during the autumn, then finally closing down and sleeping before the process starts all over again.

Blurring over details, we put Plato as one with, or a predecessor of, paganism as we read it, and the key philosophical thinker. He never believed in a Creator God, but in one of his dialogues, the *Timaeus*, he did argue that there is a deity of some sort, the Demiurge, who put everything in order and who gave to the earth its living spirit or soul. The divine designer realized that the intelligent is better than the unintelligent and that the physical on its own cannot supply this, Plato explained, and so endowed the world with a soul: "Guided by this reasoning, he put intelligence in soul, and soul in body, and so he constructed the universe. He wanted to produce a piece of work that would be as excellent and supreme as its nature would allow. This, then, in keeping with our likely account, is how we must say divine providence brought our world into being as a truly living thing, endowed with soul and intelligence," Plato wrote. Note that the use of the word "soul" here—and this seems to be general practice in paganism—is not strictly that of the later Christian "soul," a kind of mind that can exist after death. It is more an animating force, the life principle, and so is not necessarily (although it could be) conscious. It is not itself a Creator God or a designer God, but it is something that one could (and some would) worship, and it is that which gives the earth value. Life is good. Plato's world picture is one that incorporates a respect for Mother Earth. She is there to offer her bounties to humans, but they in return have the obligation to look after her and make sure she is not despoiled.

There were those who tried to fit Plato's world soul into the Christian scheme of things. In the fourth century, Calcidius helpfully gave a ready solution. The Demiurge is God the Father. The basic archetypes on which all being is based and modeled—what Plato called the "Forms" or the "Ideas"—collectively make up God the Son. And neatly, the world spirit that pervades all physical reality is God the Holy Ghost. But ultimately for the Christian it had to be Jesus first and Plato second. It isn't until after the Enlightenment and the consequent weakening of Christianity on the cultural psyche of western Europe that world souls could really start to make a fresh stand. The Romantics and their followers were great enthusiasts. Thus the poet William Wordsworth wrote about nature in his great poem "Tintern Abbey":

> And I have felt
> A presence that disturbs me with the joy
> Of elevated thoughts; a sense sublime
> Of something far more deeply interfused,
> Whose dwelling is the light of setting suns,
> And the round ocean and the living air,
> And the blue sky, and in the mind of man:
> A motion and a spirit, that impels
> All thinking things, all objects of all thought,
> And rolls through all things.

The New England transcendentalists—men like Ralph Waldo Emerson and Henry Thoreau—found world-soul thinking a welcome alternative to the dogmatic Calvinism that dominated the pulpits of their day. Especially in the case of Thoreau, this was bound up with a general love of nature and urges to go out and find and live in and cherish the wilderness. And so it continued into the twentieth century. The naturalist Aldo Leopold, long a worker for game and land management out of the University of Wisconsin, is probably the only person who has become as beloved by American environmentalists as Thoreau. *A Sand County Almanac*, his work published posthumously in 1949, made him a sage of the modern Green movement. The earth-as-an-organism notion lies at the heart of his thinking. Many, he claimed in an early essay (1923), "have felt intuitively that there existed between man and the earth a closer and deeper relation than would necessarily follow the mechanistic conception of the earth as our physical provider and abiding place." He added, "Philosophy, then, suggests one reason why we cannot destroy the earth with moral impunity; namely, that the 'dead' earth is an organism possessing a certain kind and degree of life, which we intuitively respect as such."

Similar thinking pervaded the writings of others, notably Rachel Carson and *Silent Spring*, her manifesto from 1961 against the chemicals polluting the environment that led to the banning of DDT in the United States and helped to inspire a regime of environmental-protection legislation in North America and western Europe. *Silent Spring* shrieked out that the earth is an organism. This is the way Carson saw the world—its soil, its waters, its weather, and its plant, animal, and human inhabitants. It is all interrelated. It is one, not many. If humans start to disturb one part of the world, she wrote, then it resonates throughout the rest of the world. Her language pointed to this: "Earth's Green Mantel," to quote one chapter heading; "Nature Fights Back," to quote another. And all is integrated. Carson wrote of "the obligation to endure." Of dying squirrels, brought to their state by pesticides, she wrote, "By acquiescing in an act that can cause such suffering to a living creature, who among us is not diminished as a human being?" And again and again it is stressed that humans are part of the picture, suffering with the others, and there is the whole analogy between the way people are wrecking the earth and the way they are wrecking their bodies.

Yet, prominent though this kind of thinking clearly was, perhaps no one quite anticipated the shock and subsequent controversy when the British physical chemist James Lovelock around 1970 announced his Gaia hypothesis, writing in *Homage to Gaia* (2000), that the earth is homeostatic and hence properly considered an organism: "As Pasteur and others have said, 'Chance favours the prepared mind.' My mind was well prepared emotionally and scientifically and it dawned on me that somehow life was regulating climate as well as chemistry. Suddenly the image of the Earth as a living organism able to regulate its temperature and chemistry at a comfortable steady state emerged in my mind. At such moments, there is no time or place for such niceties as the qualification 'of course it is not alive—it merely behaves as if it were.'"

Supported by the American cell biologist Lynn Margulis, excoriated by the scientific community (led by Richard

Dawkins), and loved by the general public, Gaia quickly became a part of public consciousness. And it was taken up, above all, by the Neo-Pagans. The Californian, self-styled Neo-Pagan Oberon Zell-Ravenheart openly endorses an organismic view of reality, seeing all as interconnected. In an interview in 2011 with one of the authors (Ruse), he said: "As Cicero said: '*Omnia vivunt, omnia inter se conexa.* Everything is alive; everything is interconnected.' This is what I consider the core of the Ancient Wisdom." Following an intense mystical experience, as he tells us in a collection of pieces from a Pagan magazine, *Green Egg Omelette* (2009), Zell-Ravenheart came to realize that "it is a biological fact that all life on earth comprises one single living organism! Literally, we are all 'One.'" He continued, "The blue whale and the redwood tree are not the largest living organisms on Earth; the entire planetary biosphere is." The parts of the whole have the same relationship to the whole as do the parts of the body, as one finds in the case of humans. Start messing around with the parts or removing them, and it has consequences for the whole. He added, "You can't kill all the bison in North America, import rabbits to Australia, cut or burn off whole forests, or plow and plant the Great Plains with wheat without seriously disrupting the ecology. Remember the dust bowl? Australia's plague of rabbits? Mississippi basin floods? The present drought in the Southwestern U.S.?" The world "is a single living organism, and its parts are not to be removed, replaced, or rearranged," Zell-Ravenheart wrote.

As strange as this may sound to most modern people, historically, Zell-Ravenheart's spiritual roots are as deep as those of conventional Christians; indeed they are as deep as those of any religious belief system. Moreover, Gaia promotes as strong a call for environmental action as anything to emerge from more conventional religions. Gaia enthusiast, philosopher Mary Midgley in an essay "The Unity of Life" (2005), spoke for many environmentalists when she wrote, "Man needs to form part of a whole much greater than himself, one in which other members excel him in innumerable ways. He is adapted to live in one. Without it, he feels imprisoned; the lid of the ego presses down on him." Individuals must transcend the personal and think of the group. Against the kind of thinking she finds in much of science, particularly the thinking of the Darwinian, she wrote, "The metaphysical idea that only individuals are real entities is still present in this picture and it is always misleading. Wholes and parts are equally real." And she added, "Like babies, we are tiny, vulnerable, dependent organisms, owing our lives to a tremendous whole."

ORGANISMIC BIOLOGY

The Gaia hypothesis hovers in the borderland between science and religion, standing in stark contrast with modern developments in science. The trend in biology, especially evolutionary biology, has been reductionistic—breaking things down into parts and explaining from the bottom up—and it has been very successful too. Species good, organisms better; organisms better, genes best. Richard Dawkins particularly holds strongly against the idea that the world might in some sense be an organic whole, thus giving value to its parts, which include plants and animals. It is all a matter of natural selection, and that means that animals and plants look to their own short-term interests—those being survival and reproduction. They care not at all for the welfare of others, especially not the welfare of groups. He said in the lecture "Sustainability Doesn't Come Naturally: A Darwinian Perspective On Values" (2001), "All animals do what natural selection programmed their ancestors to do, which is to look after the short-term interest of themselves and their close family, cronies and allies." Dawkins was not particularly impressed with the idea that anyone or anything does anything simply because it is sensible or right in some way. "In the 1950s when it was becoming fashionable to worry about over-population and pollution, ecologists talked about prudent predators," he added. "They thought wild predators didn't over-hunt their prey." Unfortunately that just isn't so, or at least hardheaded biologists and ecologists no longer think it is true. Organisms never think in terms of the group. It is always the individual. Those organisms that didn't think this way were wiped out by those organisms that did think this way. There really is no design or purpose. Just "blind, pitiless indifference," Dawkins and his ilk said. Biology is all about "selfish genes," maximizing their own rewards.

The pagan (or fellow traveler) takes a stand against this kind of thinking. He or she endorses a "holistic" view of the world, seeing everything as an integrated whole and decrying reductionism as missing the big picture. Analogously, it might well be that a Gaia hypothesis, whether seen as metaphorical or literal—and paradoxically without necessarily abandoning the Dawkins selfish-gene perspective on things—could push science to look at interconnectedness in very profitable ways. For instance, a key compound impacting global warming is one called dimethyl sulfide (DMS). It is produced by phytoplankton, photosynthetic microbes (algae) drifting in sunlit seawater. The phytoplankton also get taken up into the marine atmosphere over the ocean, there producing DMS, which is oxidized into sulfuric acid, which in turn induces the formation of clouds. Inasmuch as clouds reflect sunlight back into space, DMS is crucial in reflecting the sun's warming rays from reaching the earth. The big problem is that acidification of the oceans cuts down on phytoplankton production, which can lead to fewer clouds and more global warming. Yet in the 1980s no one understood precisely how the overall system worked.

As it just so happens, the reductionist-minded biologist William Hamilton, who had inspired Dawkins's thinking of selfish genes during the 1970s, got interested in the DMS problem during the 1990s, very much in the context of Gaia thinking. This does not mean Hamilton dropped his science—very far from it—but he acknowledged Lovelock's essential point that if scientists are to make progress on the problem, they must think in terms of the whole picture and not just parts. This he did, in the essay "Spora and Gaia: How Microbes Fly with Their Clouds," appearing in 2005 and coauthored by a young scientist Tim Lenton, showing how even if the microbes being lifted up might not benefit, genetically identical clones might benefit and so selfish-gene selection can take over. "Especially on sunny days with their high midday peak of radicals in the air, just when local convection is strongest, there seems a good chance for some situations in which convection due to algal DMS emissions generate a local increase of wind speed in a matter of hours," Hamilton concluded. This opens the way for natural selection. "If white tops are augmented or initiated by this increase, then the take-off process already described can potentially pay back to DMS-emitting algae at the individual or clonal patch level of selection (or to individuals via inclusive fitness) an extra possibility for causative genes to become airborne, and to disperse rapidly away." Clouds now get involved, both to protect the microbes (algae), especially from such things as ultraviolet radiation, and to transport them away from home to new grounds (or seas, rather). In other words, the algae have a motive to create and to promote clouds. Even if the actual algae doing this work die, the selective benefits accrue to their genetically identical relatives.

Hamilton stays within his reductionist paradigm even as he explains holistic interconnections in natural processes. Many would say that he made the paradigm, if that reductionist paradigm is taken to mean that natural selection only benefits the individual, where here the individual might include family members or clones. For an even more dramatic example of Gaia-type holistic thinking influencing an otherwise reductionist-minded evolutionary biologist and prominent proponent of Hamilton's selfish-gene hypothesis, consider E. O. Wilson. He is as much a non-believer as Dawkins; but, as the end product of a chain that goes back to the American transcendentalists (and beyond them to the German Romantics), Wilson sees all of life as an interconnected whole. This is expressed through his "biophilia" hypothesis, in the book of that name, *Biophilia* (2004). "To explore and affiliate with life is a deep and complicated process in mental development," he wrote. "To an extent still undervalued in philosophy and religion, our existence depends on this propensity, our spirit is woven from it, hope rises on its currents." Individual organisms, individual species, are part of a larger network, and no one or no group can take itself out of the whole and live apart in isolation.

In an interesting way, because of his holism, seeing all as interconnected, Wilson is uncomfortable in picking out global warming as such (as we have done thus far in this chapter) as the dominant issue. Global warming is a vitally important issue, but it must be placed in the context of all issues of environmentalism, human and nonhuman. For this reason, Wilson wants to take the discussion up to the "metalevel," thinking not just about what people should do, but why they should do what they should do. He thinks that what has evolved is in some sense good. Because humans are the top of the heap, Wilson affirms, they are in some sense the supreme good, and moral action should be directed to the welfare of humankind. Yet, showing that although his Christian childhood is still at work he has not really flipped back to it, Wilson sees that this anthropocentricism must be combined with a sense of the holistic, organic nature of things. Humans, he said in an interview at Williams College (1990), should tackle the paradox of "the two themes that form the fundamental basis of ethics, the expanding-circle theme that gives rights to all species, versus the anthropocentric theme that measures all good in the coin of human welfare." This will be difficult, Wilson conceded, but is not impossible. "The two are resolved in part by noting that for human survival and mental health and fulfillment we need the natural setting in which the human mind almost certainly evolved and in which culture has developed over these millions of years of evolution. Perhaps both of those arguments can be joined to create the prudence concerning the environment and our own populations that is so desperately needed," he reasoned.

Note then that for Wilson, morality is natural. It is something that emerges from the ways that things are, and hence morality about the environment and the future of the planet is going to emerge very much from the nature of the environment (including animals and plants) and the things pertinent to the future of the planet. In this Wilson is very much at one with the pagans. For them, nature dictates what people should think and do. The same is very much true for Wilson. Note also that although he gives humans a special status, he is with the pagans (and the transcendentalists) in seeing life as interconnected. Wilson thinks that biology—natural selection—can promote group interests, and he sees it at work within collections of organisms like the human species. But he perceives something wider. It is not just a matter of human welfare. Unless they pay attention to the welfare of others and of nature, humans will suffer and wither and die.

Wilson stresses always that it is not just that humans directly benefit from nature—such as by the medicines from tropical plants—but also that there is a broader, more fundamental benefit or need. Quite literally, he maintains, humans could not survive in an artificial world. As they have co-evolved with the natural world, they need the living environment to flourish. "I doubt that most people with short-term thinking love the natural world enough to save it," Wilson conceded in an interview on *Nova* in 2008. "But more and more [people] are beginning to get a different perspective, particularly in industrialized countries. It's becoming part of the culture to think rationally about saving the natural world. Both because it's the *right* thing to do—and notice the quick spread of this attitude through the evangelical community—but we will save the natural world in order to save ourselves." Part and parcel of this "is what can only be broadly called 'the love of nature.' I think that an attraction for natural environments is so basic that most people will understand it right away," he added. Seeing this, and grasping for every reed that might save civilization, Wilson reached out to evangelical Christians with his book from 2006, *The Creation: An Appeal to Save Life on Earth*. It became a national best seller.

For Wilson, therefore, fighting global warming, caring about life generally, promoting sustainability, are indeed matters of self-interest. But he would be uncomfortable with suggesting that the tools to do this are at most serendipitous by-products of selection working for other ends. He would see biology promoting more general concerns and interests. And as part of this overall picture, he would see that promoting human interests is necessarily promoting the interests of all—at least this is so if people are to function well and to be successful in their aims. "The living environment is what really sustains us. The living environment creates the soil, creates most of the atmosphere. It's not just something 'out there.' The biosphere is a membrane, a very thin membrane of living organism. We were born in it, and it presents exactly the right conditions for our lives, including psychological and spiritual benefits," Wilson said in the *Nova* interview, bringing one extreme of scientific naturalism almost full circle to deep-green spiritualism.

Wilson has very great respect for Pope Francis, but the two men could not be further apart on their fundamental presuppositions. And that is a good point on which to bring this discussion to an end. The inhabitants of this earth face serious physical and social issues. Standing still and doing nothing is not an option. Hard thinking about the science and technology combined with deep moral seriousness and the religious conviction of believers are absolute requirements. Together with the realization that others, no less learned and no less serious, will come from other directions. No one should feel threatened by differences, nor should anyone quake and yield because there are differences. But if humans are in this together, sympathy and understanding are essential. Then perhaps we can move forward together.

Bibliographic Essay: Where to Look from Here

With Pope Francis reaching out to scientists, and scientists like E. O. Wilson reaching out to religious believers in shared concern over the environment, in a sense this book has come full circle. It began by describing the complementary interaction of Greek natural philosophy and Catholic theology during the medieval period, with each supporting the other in a view of the cosmos that had the earth (and therefore humanity) at its center and the heavens (and therefore God) at its all-encompassing periphery. The Copernican Revolution shattered this medieval cosmos, and, as the status of science grew, an ever more complex relationship developed between science and religion in the West. Episodes of conflict emerged even as areas of complementarity remained, while a wary coexistence seemed to characterize the general norm. Matters have been different in the East, but even there the relationship is complex, with science and religion seeming to co-occur with each other more than they coexist, conflict, or complement. Nowhere, however, can the relationship simply be dismissed as two nonoverlapping magisteriums with separate visions for humans and the universe. Because of their claims of authority, science and religion inevitably overlap. And because their followers see them as having external foundations in nature or the supernatural, rather than simply human innovations, both have certain contested claims to "truth." No single book has told this entire story. At most, this one seeks to introduce the topic by raising a series of episodes in the history and philosophy of science. Seeing it as a beginning rather than an end, we want to close by suggesting some further readings that we have found helpful or inspiring in our work and in drafting this book. But even this list is far from comprehensive.

INTRODUCTION: WHAT'S THE FUSS?

Digging into the science-and-religion relationship requires some understanding of the nature of both science and religion. Two classic works on science, already over fifty years old but still worth reading, are Karl Popper's collection of essays, *Conjectures and Refutations*, and Thomas Kuhn's *The Structure of Scientific Revolutions*. Both men take science very seriously indeed, but whereas Popper tends to see it as some striving for objective truth—what he felicitously calls "science without a knower," meaning that the nature of the person producing the science is irrelevant to its truth—Kuhn takes a more sociological approach, and his famous notion of a "paradigm" is very much something that appealed to the scientists and their communities.

There are countless works on the nature of religion. Part of the challenge here is that different religions tend to give different perspectives on the nature of the field. Very popular recently have been the writings of the literary scholar and children's novelist C. S. Lewis, and his *Mere Christianity* is something that many would recommend as an introduction to that religion. Whether all of his co-religionists would agree with this, however, and what might be the feelings of those out beyond Christianity is a moot point. This perhaps is the idea we would want to stress above all others—that religion is such a broad-ranging phenomenon that generalizations are almost always misleading, and that this is particularly so when one raises one's gaze to traditions that are not one's own.

Moving directly to works that look at the relationship between science and religion, a classic still worth reading is the book that came out of Ian Barbour's Gifford Lectures, *Religion and Science*. In those lectures, he introduces his fourfold method of analyzing the science-religion relationship, going all the way from outright warfare to total harmony and integration. There have been criticisms of the categories, but they are a starting point for further work. One of us, Michael Ruse, has also made a contribution to this discussion. His *Science and Spirituality*—a book that is a lot less pompous than the title suggests, is an attempt to view the science-religion relationship through the lens of metaphor and how it functions in the very best quality science. Stephen Jay Gould's *Rocks of Ages* introduces his controversial idea of a "magisterium," a kind of world view or perspective. What many found unsatisfactory about the book was less this idea and more the use he made of it—upsetting believers by achieving harmony in reducing religion to moral sentiments, and upsetting nonbelievers by trying in the first place to find harmony between science and religion.

Three collections of essays introducing the complexity approach to the study of science and religion, and refuting the earlier conflict model, are *God and Nature* and *When Science and Christianity Meet*, both edited by University of Wisconsin historians of science David Lindberg and Ronald Numbers, and *Science and Religion around the World*, edited by Numbers and John Hedley Brooke. For single-author books taking much the same approach, see Brooke's *Science and Religion* and Peter Harrison's *The Territories of Science and Religion*. For an introduction to the history of science and Eastern Christianity, which is largely absent from the books listed above, a good place to start is the 2016 article by Efthymios Nicolaidis and others, "Science and Orthodox Christianity: An Overview," in volume 107 of the journal *Isis* (the journal of the History of Science Society) plus the published responses in *Isis* to that article by Numbers, Brooke, Harrison, and others. And for the classic work on the other side, try to stay awake through Andrew Dickson White's two-volume *A History of the Warfare of Science with Theology in Christendom* (1896).

CHAPTER 1. LOOKING UP TO GOD OR THE COSMOS

Theorizing about the heavens starts with the Greeks not the Jews, so look to the former for insights. *Creationism and Its Critics in Antiquity*—thinking in a pagan context obviously, not a Jewish or Christian one—by David Sedley is a compelling read. Before he wrote *The Structure of Scientific Revolutions*, Thomas Kuhn wrote *The Copernican Revolution*. Although now somewhat dated by subsequent scholarship, it remains a classic read, and thanks to the huge number of excellent illustrations, the technical details have never been made easier to understand. In line with his overall philosophy, Kuhn stresses that historians must judge ideas on their merits as seen in their day. Thus for instance, as judged by history, simplistic suggestions that the Scientific Revolution meant a move from God are just plain false. Even the Galileo story is more nuanced than critics of religion would have it. Richard Westfall's *Science and Religion in Seventeenth-Century England* delivers just what the title offers. Westfall has a huge grasp of the range of science and at the same time brings a full understanding and appreciation of the nature of religion and of its importance. For his part, Robert S. Westman set the record straight on Kuhn's book in his article "Two Cultures or One?: A Second Look at Kuhn's *The Copernican Revolution*," published in volume 85 of the journal *Isis* in 1994.

CHAPTER 2. THE TAO OF PHYSICS AND OTHER BIG IDEAS

Historians Peter J. Bowler and Iwan Rhys Morus offer a good overview in their *Making Modern Science*. In the nineteenth century, many scientists—including physical scientists—were deeply religious, and no historian does a better job untangling the scientific contributions and religious views of one of them than Geoffrey Cantor in *Michael Faraday: Sandemanian and Scientist*. Such religiosity among physical scientists was less common in the twentieth century. Arthur Eddington, living at the beginning of the twentieth century, was a leading astrophysicist—it was he who tested Einstein's theory of relativity—and a lifelong Quaker. In a talk he gave to his co-religionists, Eddington spelled out how he sees the relationship between science and religion. Given in 1929, his lecture *Science and the Unseen World* is still worth reading, as is Matthew Stanley's biography of Eddington, *Practical Mystic*. Moving toward the present, *The First Three Minutes* by Nobel Prize–winner Steven Weinberg is a hugely interesting book about the Big Bang. Whether this shows, as he claims, that there is no meaning, and hence no god, is for readers to decide. Supposedly everybody bought but few may have finished Stephen Hawking's *A Brief History of Time*, a work that deals with many of the same issues, although at times more overtly playing on science-religion themes in ways that seem designed to upset believers. Religious believers and seekers typically prefer books like *God and the New Physics* by English physicist Paul Davies.

CHAPTER 3. THE BRAIN, THE MIND, AND THE SOUL

Steven Pinker is a Harvard psychologist and one of the very best popular writers about science. His book *How the Mind Works* is a simply dazzling account of how people think rationally, why so often they do not think rationally, what drives people, why they are social, and so much more. It provides a terrific foundation to think about the relevance of what is known about the mind to the religious context. For a look at the late-nineteenth- and early-twentieth-century foundations for the modern scientific understanding of the mind and soul, we recommend *The End of the Soul* by Jennifer Michael Hecht. Of course, people have been wrestling with these issues since ancient times, when the Greeks and the Jews flourished. For some great classics of the past, then, the Oxford World's Classics edition of *The Confessions of Saint Augustine* should not be missed, and the same is true of a work written over a thousand years later, *Discourse on Method* by the French philosopher René Descartes. You can read how very different after the Scientific Revolution is the thinking of Descartes—far more mechanistic—even though he was no less a sincere Christian than Augustine. Jumping to the present, trying to make sense of Christianity in an age of evolution and of how people should now think of the mind and its doppelgänger the soul, the eminent theologian Nancey Murphy does a sterling job in *Religion and Science.*

CHAPTER 4. ROCK, FOSSIL, GOD

A wonderful background account of the theological worries posed by geology in the nineteenth century is the reissued 1996 edition of *Genesis and Geology* by Charles Gillispie, with a foreword by Nicolass Rupke updating the material. Although it is now well over half a century old, it is still a thrilling and informative read. Anything written by Martin Rudwick is worth the price of the book. *Earth's Deep History* synthesizes a huge amount of his earlier scholarship. An excellent book on Cuvier is William Coleman's *Georges Cuvier, Zoologist,* and an equally excellent book about Lamarck is Richard Burkhardt's *The Spirit of System.* Rounding this out, a truly superb book on Darwin and geology, bringing in his religious views, is Sandra Herbert's *Charles Darwin, Geologist.* We mentioned another book by Gould earlier, but for a lively tour of paleontological discovery focused on the Precambrian Burgess Shale and its broad implications, it is hard to beat Stephen Jay Gould's engaging *Wonderful Life*, which can be profitably paired with Simon Conway Morris's *The Crucible of Creation*.

CHAPTER 5. DARWINISM AND BELIEF

There is so much material here it is hard to know where to start, so let us begin with a couple of our own books! Edward J. Larson's *Evolution* offers a broad and easily understood introduction to the entire history of evolution theory, with due attention to the religious, political, and social resistance it has encountered. Michael Ruse's *Can a Darwinian Be a Christian?* is right at the other end of the pole. It is not a work of history but more of philosophy and theology, going through the potential points of conflict like free will and determinism. The author concludes that it is possible to be a Darwinian and a Christian, but it isn't easy. John F. Haught's *God After Darwin* is a thoughtful book by a Catholic; Alister McGrath's *Darwinism and the Divine* is a good Protestant take on things; and C. Mackenzie Brown's 2012 book, *Hindu Perspectives on Evolution*, is the best we've seen from an Eastern religious perspective. For top contemporary evolutionary biologists confessing their Christian beliefs, we recommend Kenneth Miller's *Finding Darwin's God* and U.S. National Institutes of Health Director Francis Collins's *The Language of God*. And anyone who has not read Darwin's *On the Origin of Species* should try. It is remarkably readable. And for reasons to throw our book away unread, don't miss Richard Dawkins's best seller, *The God Delusion*. He writes with a Fundamentalist's fury and regards people like us, who see virtue in fostering a relationship between science and religion, as beyond the pale. In *The Creationists*, historian Ronald Numbers relates how, during the twentieth century, a small band of American biblical literalists created a worldwide movement devoted to giving scientific-sounding support to the Genesis account of cosmic, geologic, and organic creation. To hear this from the movement's own Aaron and Moses, read the 1961 creationist classic *The Genesis Flood* by John Whitcomb and Henry Morris.

CHAPTER 6. THE EVOLUTION OF HUMANITY

The study of human evolution—paleoanthropology—is a tremendously fast-moving field, and anything said today stands the risk of being out of date tomorrow. A good if somewhat dry account of the history of the search for human origins is Peter Bowler's *Theories of Human Evolution*. More exciting, and written by one of the leading researchers in the field, is *The Fossil Chronicles* by brain scientist Dean Falk. She shows with devastating clarity the extent to which prior expectations and prejudices color the work that is done on human evolution. And to read about how popular resistance to the idea that humans come from monkeys played out in court, Edward J. Larson's Pulitzer Prize–winning *Summer for the Gods* (the author is too modest to mention its distinction but his coauthor has insisted) is about the Scopes Monkey Trial in 1925, when a young schoolteacher in Tennessee was prosecuted for teaching that humans are descended from apes. It is a great story but also deeply revealing about the reasons why the science-religion debate can get so heated and why social and cultural issues, rather than anything strictly theological, are often driving the agenda. For a discussion of the battles over hominid fossils, we recommend *Bones of Contention* by British science writer Roger Lewin as a good place to start.

To bring this story of religious responses to evolution up to date, we should note that Pope John Paul II was doctrinally very conservative, but when it came to science—perhaps in part a reflection that he had been a professor at Jagiellonian University in Poland, where his most famous predecessor was none other than Nicolaus Copernicus—he was much more open-minded, endorsing evolution, Darwinian evolution, and even the evolution of humans. However, as his papal letter of 1996 shows, he drew the line at a naturalistic origin of the human soul. Here God steps in. For a sense of how fragile from a religious perspective is discussion of these matters, look at the article by the Protestant theologian John Schneider, "Recent Genetic Science and Christian Theology on Human Origins: An 'Aesthetic Supralapsarianism,'" published in 2010 in the journal *Perspectives on Science and Christian Faith*. For suggesting that modern science throws doubt on the historical authenticity of Adam and Eve, he lost his job at the Reformed Church–affiliated Calvin College. The science-religion debate is more than just history.

Many of the best books about human evolution are by top researchers in the field; who would have guessed that they could write so well and do extraordinary science? Simply one of the all-time best books of popular science is by Donald Johanson (a paleobiologist) and Maitland

Edey (a science writer), *Lucy*, all about the little australopithecine that so stimulated the search for human origins. Edward O. Wilson's Pulitzer Prize-winning *On Human Nature* is a controversial approach to understanding humankind, including its sociality, its morality, its religions, from a Darwinian perspective. And to know something about the exciting new field of ancient DNA research, *Neanderthal Man: In Search of Lost Genomes* by Svante Pääbo is the place to start. All of these books raise religious issues, if not directly.

CHAPTER 7. SEX AND GENDER

Sigmund Freud is not really that fashionable these days—we suggest that he got some things wrong (or at least backward), and he was not without prejudice—but really it was he who smashed through two thousand years of thinking about sexuality. His writing is remarkably readable. His *Three Essays on the Theory of Sexuality* and *The Future of an Illusion* show how a totally secular approach to these issues challenges the Judeo-Christian approach. More recently, a controversial history but with many deep insights is that of the French philosopher-historian Michel Foucault, *The History of Sexuality* (especially volume 1). Moving on to gender and science, among the many starting points are Evelyn Fox Keller's *Reflections on Gender and Science*, *Nature's Body* by Londa Schiebinger, and Carolyn Merchant's *The Death of Nature.*

When it comes to reading about religious thinking on sexuality, there are myriad options. One of the authors of this book (Ruse) came from a Quaker background and found deeply inspirational an essay written some fifty years ago, "Towards a Quaker View of Sex." It is dated but offers a truly spiritual account of how religious believers can resolve conflicts between their religion and the ongoing findings of modern science. Looking to other religions, we recommend starting with the collection *Buddhism, Sexuality, and Gender* edited by José Ignacio Cabezón. Follow this with *Sexuality in Islam* by Abdelwahab Bouhdiba, which is not only detailed about the theology and practices of sexuality, but argues that much of the repressive nature of modern Islam is more a function of culture and economic uncertainty than anything found in the religion itself. As noted in our text, *Reweaving the World*, edited by Irene Diamond and Gloria Orenstein, includes an interesting array of pioneering articles on feminist spirituality, gender, and deep ecology.

CHAPTER 8. EUGENICS, GENETICS, AND PLAYING GOD

In the Name of Eugenics by Yale historian Daniel Kevles is the place to begin with the history and practice of eugenics. *Sex, Race, and Science* by one of us, Edward J. Larson, deals with the story of the American South and the politics of eugenics. Follow these up with *Racial Hygiene* by Stanford historian Robert Proctor, who shows with devastating clarity how Nazism led to the perversion of a technology, including medicine.

In the modern realm, Philip Kitcher's *The Lives to Come* is a thoughtful overview of modern genetics, how it applies to humans, and what this means to us all in the future, both with respect to medical care and to improving the basic nature of humankind. There is expectedly a huge literature dealing with genetics and religion. A running start into the topic of eugenics and religion is provided by Christine Rosen's *Preaching Eugenics*, Sharon Leon's *An Image of God*, and the entries on religious topics in *The Oxford Handbook of the History of Eugenics*, an exceptionally impressive collection of essays edited by Alison Bashford and Philippa Levine. In respects, the writers of fiction are as challenging and thought-provoking on this topic as any. Typically, Kurt Vonnegut's *Galápagos* is slightly crazy, but he does set readers thinking about the effects of the genes and what people would really want if they had a totally free hand in these matters. As always, life is not quite as simple as it seems at first.

CHAPTER 9. LIVING ON EARTH

By all means start with the Roman Catholic encyclical on the environment, *Laudato Si'*, by Pope Francis, and *Encyclical on Climate Change and Inequality*, which Francis authored with an introduction by Harvard University historian of science Naomi Oreskes in 2015. These works provide a good perspective not only on many of the challenges life on earth is facing but also on a Christian response. Follow this up with a work on Christianity and global warming, *A New Climate for Theology* by the distinguished theologian Sallie McFague. For a broader, more historical view that picks up on some of the issues that excite and interest the pagans, then look at *The Gaia Hypothesis* by one of us, Michael Ruse. An interesting attempt to link Eastern thinking with some of the insights of the pagan movement, particularly as they find their way into so-called "deep ecology"—a movement that sees value in nature rather than value as something to be read into nature—is *Buddhism and Deep Ecology* by Daniel H. Henning. A useful book is *Buddhism and Ecology* edited by Mary Evelyn Tucker and Duncan Ryuken Williams. Really though, serious discussion is only now beginning on so many of these issues. This is a major reason why we wrote this book, to get others engaged and working on the problem. If you have as much fun reading it as we have had writing it, you will be well rewarded and we will be well satisfied.

Index